T0176197

BEYOND
9 TO 5

Maps of the Mind
Steven Rose, General Editor

Maps of the Mind
Steven Rose, General Editor

BEYOND 9 TO 5

Your Life in Time

SARAH NORGATE

Columbia University Press
NEW YORK

Columbia University Press
Publishers Since 1893
New York   Chichester, West Sussex

Copyright © 2006 Sarah Norgate
All rights reserved

Library of Congress Cataloging-in-Publication Data
A complete CIP record is available from the Library
of Congress.
ISBN 0-231-14008-8
♾

Columbia University Press books are printed on permanent
and durable acid-free paper.

Printed in the United States of America

c 10 9 8 7 6 5 4 3 2 1

First published by Weidenfeld & Nicolson Ltd., London

For my Grandmothers, Betty and Connie

CONTENTS

ACKNOWLEDGMENTS

To the friends and colleagues who travelled with me during this venture, I want to say thank you for staying with me until 'the end'. At Orion, in particular I wish to thank Ian Drury and Ilona Jasiewicz for their guidance throughout all stages of preparation. I must thank Steven Rose for inviting me to contribute to the 'Maps of the Mind' series, and say that I have appreciated his capacity to be able to stand in the shoes of a first-time author, and be able to provide a map of this unfamiliar territory.

Some of the ideas for this book arose from conversations about time from the 'Mind and Brain' seminar series held at The Open University in 1999, others from my research interest in the role of temporal information in child development. For this reason, I wish to thank the Trustees of The Mary Kitzinger Trust, UK, for enabling me to hatch this line of research back in 1992 (some of which is reported in Chapter 7).

To those who shared ideas, commented on drafts, helped with motivation and were patient with me, their generosity is deeply appreciated. Special mention must go to Gavin Bremner whose catchphrase, 'carry on', was a great calming influence during the writing, and who remained tolerant of our 'lost time'. Also to Coralie Farmer, Niall Adams and Rachel Hutchison for their brutal honesty and their generosity. Thanks to other minds (and motivators): Mary-Jane Kehily, Richie Joiner, Sir Roger Bannister, Xavier Hassan, Eileen Mansfield, Sven Brautigam, Angus Gellatly, Harry Heft, Fred Toates, Kevin McConway, Ken Richardson, Kier Thorpe, Jay Joseph, Becky Shaw, Dipesh Chaudhury, Fiona Dugdale, Ali Kirkbright, Heather Montgomery, Matt Schlanker, Paula Piggott, Sarah Welch, Jackie Rossi, Nige Vincent, Birendra Giri, Erica Morris and Adam Sumner. Thanks to my family for their interest in this book and its progress.

BEYOND
9 TO 5

INTRODUCTION

You now have seven minutes to read as much of this as you can. Go! So why insist on a countdown? Apparently, seven minutes is about the average amount of time adults spend book-reading each day. If this is rather less than you expected, note that it excludes time spent flicking through magazines or moments spent craning your neck over someone else's shoulder in an effort to catch the news headlines.

Despite it being the most logical next step, it would be rather thoughtless of me to press ahead with a comprehensive breakdown of how we typically use the remaining twenty-three hours and fifty-three minutes. Aside from the experience being mind-numbing for all of us, it would start this book off heavily invested in a single relationship with time – and how much we allocate to different activities – when actually this book's purpose is not to focus on that. Although the subject of how we carve up our days does enter into the first chapter, this book's ultimate quest is to unveil the multiple relationships we have with time in our everyday life.

Plenty of titles already exist about the subject of absolute time, be it in terms of the billions of years of cosmologists or the nanoseconds of particle physicists. Instead, in this book, I immerse us in the psychology and some of the biology behind our *lived* time – after all, as we shall see, there are some enticing reasons to become familiar with the range of our various relationships with time.

Most of us have particular default modes with time. If you aren't yet in touch with yours, just think 'time' and see what springs to mind. The nine to five routine? Constantly feeling in a rush? Never having enough? Juggling commitments? According to the research into the psychology of time, these particular aspects only touch on some of our relationships with it. If, say, nine to five carries negative connotations for you, then hopefully after

1

reading this book you may see you have the liberty to get beyond this mode of thinking and prioritise one of your other relationships with time.

Time and space are inextricably linked, yet certain imbalances emerge between them in terms of the recognition they receive in society. There is no doubt that time gets plenty of recognition in conversation. After all, we use the word fifteen times more frequently than we do 'space'. Then there are topical time-related issues like work–life balance and European working time directives, which rarely leave the media spotlight. Nevertheless, compared to the status physical space has achieved in westernised society, this recognition remains essentially talk and legislation. You only have to look at people's job titles to see that they routinely reflect our engagement with different sorts of everyday physical space – take architects, cartographers, urban planners, estate agents, groundskeepers, geospatial consultants, interior designers and landscape artists. According to the *Standard Occupational Classification*, job titles exclusively reflecting our relationship with time are less common. A timekeeper in sporting events will primarily deal with the immediate duration of the event. A waiting-time coordinator working in a hospital may be routinely concerned with timetabling, planning and organisation over a number of months. A number of professions deal with the longer-term future or past – historians, campaigners for sustainable living and various types of forecaster. There is an ever-increasing army of gurus in time management – not to mention life coaches. Crucially though, none of these professions addresses the *multiple* relationships we share with time.

Futurists making predictions about life later in the twenty-first century also champion physical space over temporal. Their passion lies in talk about innovation in the physical domain – the possibility of lunar housing, neural meshing between humans and computers, a nanotechnological revolution, transforming urban housing, the wide availability of robots and the potential to control the rise of sea-levels. Though there are claims to a future of time travel, these remain fanciful in the eyes of most.

Consider people whose fame is based on having conquered exceptional physical territory; they achieve greater recognition than those who have survived an exceptional time. Many of us are aware that Neil Armstrong was the first person to walk on the moon. In contrast, unless you study longevity, the name of the world's officially recognised longest-lived person – Jeanne

Calment – who died in 1997 aged 122, is probably unfamiliar. So far, no one I have asked has heard of her.

So why should we put our relationship with time higher up on society's agenda? Each of this book's chapters reveals a different answer to that question, depending on which relationship with time is discussed.

Perhaps you are generally rushed off your feet, behind schedule and torn between commitments. Or maybe you would position yourself nearer the other end of the spectrum, being fairly laid-back, with no fixed schedule and comfortable passing between activities. Either way, the first chapter will give you the chance to compare your time habits with those of others around the world by looking at global trends in annual work hours, holiday allocation, pace of life and whether or not the language of saving, managing, wasting and losing time is universal. Mobile technology, communications and globalisation are taking us beyond the traditional nine to five and I look at how these are changing day-to-day living. By the chapter's close, large-scale inequalities in time use will be apparent, particularly for those who work irregular hours.

In Chapter 2, I pursue the problem of working irregular hours, showing why night work is detrimental to our health. For people who regularly stay up at night, a clash occurs between the cycle of their biological clock and the natural light–dark cycle. In examining the biology behind this clash and the mechanisms involved, we see how the characteristics of modern living – round-the-clock consumption and travel across time zones – affect our health. It is generally those who are the poorest who pick up the tab in terms of longer-term costs to their health and well-being.

As the effects of night work can go as far as impacting negatively on our longevity, it is this particular relationship with time – the time of our life – that becomes the focus in Chapter 3. At your disposal are around 2.5 billion heartbeats, a rough estimate of the number that occur during the current world average life expectancy (at birth) of sixty-five years. Whether we end up falling short or, at the other extreme, massively overshooting this allocation, we all have a 'life-long' path through time. World inequalities in life expectancy are vaster than ever before in history. The reasons some of us survive to an extraordinary age can be attributed to a combination of factors, some of which we can exert influence over to potentially extend our longevity. However, our life-long relationship with time does not only refer to life expectancy, it can also refer to how surviving family generations space themselves in time. Despite the uniquely human capacity to reflect on longevity and the spacing of new family generations, our actual ability to choose when to time new generations is far from uniform worldwide.

How long we live affects another relationship we share with time, that of our experience of it in terms of the currency of memory. A centenarian's experience of time is qualitatively different from that of a thirty-year-old. Yet, regardless of our eventual longevity, most of us appear unable to access our earliest memories. If you leaf through a selection of autobiographies you will see that a common starting point is around the age of five, rather than any younger. In Chapter 4 I show that although most of us have memories prior to this age, it is striking how few apparently merit autobiographical comment. Fortunately though, our other ways of experiencing time, like the capacity to judge its passage, stay fairly intact throughout life. In this chapter everyday examples of waiting are given, and described in relation to brain mechanisms for how we keep track of time. A central theme in this chapter is that our experiences of time are specific to the particular century we find ourselves living in. In the quest to understand how the mind experiences time, it is all too easy to assume that all human minds experience it in the same way. By drawing on selective life experiences – doing jail time, facing terminal illness, déjà vu and competing in the Olympics – I characterise the nature of our differing experiences with time.

For a fraction of our lives – as babies – we cannot talk about our experiences with time, but that doesn't mean that there are no other relationships with time getting off the ground during this early period. In Chapter 5, I look at the very beginnings of our relationships with time, which are traceable even to the foetal stage. The costs to society of babies' nocturnal antics are set against new findings from studies of sleep interventions. As well as talking about early difficulties with time, I trace the roots of other timing skills including those needed to participate in football, tennis and cricket later on in life.

The theme of skills in the timing of movement continues in Chapter 6, where I shift the focus to adulthood and unzip the nature of the millisecond-based muscular timing underlying even our most basic movements, as well as explaining the reasons behind the success of sporting superstars. With sport now firmly on the agenda for a healthy society, the more that is known about the timing of muscle movement, the better chance we have to coach tomorrow's youth towards sporting success.

Not all of us have muscle groups that perform within the same kinds of speed range as the majority of the population. Chapter 7 picks up on people's exceptional relationships with time. These unusual relationships do not have to be reflected purely in terms of talk about the speed of muscle movement, but highlight a number of conditions affected by time – Parkinson's disease, addiction, sensory impairment and autism to name but a few. In some conditions just one relationship with time has been affected, whereas in others a number are influenced. These unusual relationships with time are not always taken into account in the design of society's public spaces. For instance, some automated doors and phone services operate at fixed speeds, when there should be flexibility.

In the final chapter the various threads of our relationships with time are united. Bringing these together provides opportunities to understand further what makes us human, and to better understand links between behaviour, brain and genes, considered in the wider context of the influences of society and culture. Not only that, but it also allows us scope to question the quality of our relationships with time, and whether we want to alter anything. This includes trying to improve the lot of those who are 'time impoverished'.

Now it's time to stop the watch. If you rose to the challenge by speed reading this introduction, by reading at 300 words per minute you would have used up six minutes and two seconds of your daily allowance. So, with fifty-eight seconds left to play with, you can at least make a start on *War and Peace*.

Chapter 1

'SORRY, I HAVEN'T GOT TIME': TIME AS A PRESSING CULTURAL ISSUE

Long checkout queues, crowded station platforms, the fraught school run and overhearing other people's phone conversations are all prime occasions when you might notice time hogging the conversational limelight. Time moves around our social relationships in a mercurial fashion; from precious commodity to unfulfilled promise, it slips and slides into thoughts and actions. Listen to it taking grip of apologies, accusations, excuses, complaints, promises and the like.

'You always say you're too busy, if it was that important to you, you'd make time.'
'Time is money; let's pay someone to do the ironing.'
'Don't waste time getting caught in the rush hour.'
'Look, let's promise that when we finally do get to see each other, it'll be quality time.'

Whether it's talk about being busy, the breakneck pace of life or the work–life (im)balance, one way or another most of us put on public display our relationship with time – or rather its scarcity. Alongside this, we subconsciously soak up messages about how we 'should' relate to time as expressed through the medium of billboards, product packaging, TV, radio, magazines, newspapers and so on. Sanatogen makes vitamins for 'adults who are constantly juggling their lives'. Procter and Gamble's Wash and Go shampoo launches people out of the shower and into the next activity.

Manufacturers of the bacteria-friendly drink Yakult remind us, 'Rushing, travelling, working, stressing. How you live can upset your life.' Volvic mineral water makes sure you 'keep your natural spark going 24/7!' Other adverts persuade us that our ability to 'control' time knows no bounds. One from Orange™ tells us to 'work faster in more places', another from Vodafone Ltd says, 'Call someone; now is good', and Microsoft Office® appeals to us to 'Save time: with smarter working practices to boost productivity.' There is no shortage of ways to fast-forward our lives. Think of speed dialling, speed banking, speed dating and more recently, speed nannying*. Various products claim to maximise our efficiency by letting us juggle several activities simultaneously. Take iRobot's wire-free Roomba vacuum cleaner. You can switch it on to dash around the house, navigating under sofas and furniture, while you do something else. More traditional goods are also heavily marketed as time-friendly, like the water-resistant clock radio that lets you 'stay entertained, keep on schedule'.

With an inexhaustible array of such products to blow your cash on, there's the prospect of having more spare hours on your hands, so why not ask friends over and speed up your 'down time'? Purchase an electric tabletop wine cooler that 'chills wine fifteen times faster than the fridge'. Then do away with the chat and laughter which normally accompany the painfully slow extraction of a stubborn cork and become instead the proud owner of 'the fastest corkscrew you'll ever use'. You'll want to eat, so bolt down a ready meal together. Ensure that you maximise your potential time saving by having a microwave with a decent wattage. Zapping at 850 watts ensures that four red Thai curries and four portions of egg fried rice are served up five minutes faster than if cooked in a 750-watt model. Next, swiftly on to a round of cards – but to avoid waiting for someone to shuffle the deck use an automatic card shuffler and you'll have two packs shuffled in only three seconds. How very convenient: one evening's entertaining nicely squashed down into a fraction of the time it would normally take.

Of course, not all products claiming to meet the gold standard for time friendliness actually live up to their word. I was once seduced by an electric juice extractor, but things turned sour after a head-to-head with my older model. This only took three minutes fifty-two seconds to manually juice four

---

* Parents and prospective nannies congregate in a room and pair off.

oranges (including cleaning time) while the electric version actually clocked up a total of six minutes thirty-four seconds – including over two minutes for a major clean-up operation to remove spattered juice from the vicinity and the reconstruction of various appliance parts. If safeguarding your time is important to you then the moral of the story is clear. Squeezing oranges by hand for thirty years would free up a grand total of just over eight days for more, er, fruitful pursuits.

One appliance responsible for cutting down on – or altogether avoiding – the time devoted to preparation of food is the freezer. Back in the 1960s the home freezer was often hidden from view, banished to a garage or shed, and used to store batches of home-grown fruit and vegetables. In 1965 only 3 per cent of the UK population possessed a freezer. The opening of the first branch of Iceland in Britain in 1970 signified the arrival of another ice age to the extent that some 800 Iceland stores have since opened nationwide. The freezer is now regarded as an essential in the kitchen, and in 1995 some 96 per cent of households possessed one. More recently, the introduction of the 'frost-free' freezer has allowed people time to chill out instead of spending precious leisure time defrosting. In only a matter of minutes ready meals hop straight from the freezer to the microwave to the mouth.

Fast or 'accelerated' living or being strapped for time is not just a twenty-first- or late-twentieth-century phenomenon. Back in the mid-1800s, French commentator Alexis de Tocqueville portrayed Americans as being in a constant hurry. In the late 1880s early ads for Coca-Cola flaunted its capacity to counter a 'slowness of thought'. When Albert Einstein departed Germany on a liner bound for San Francisco in 1933 he was reported to have said that the present pace of life was too fast for the man in the street even to catch the newspaper headlines and it was imperative that we slowed down: 'A few years ago people had the chance to sit down and think. It could not be helped if some did not make use of the opportunity, but now no one is in a position to stop and think even if he desires to do so.' The 1960s saw a new crop of books about the time squeeze – again, predominantly concerning American culture – like Sebastian De Grazia's *Of Time, Work and Leisure*, which declared America a leisureless society. And bang, here we are in the twenty-first century. Déjà vu. In James Gleick's *Faster* you can read – or rather, speed read – about the acceleration of just about every aspect of daily life. An exception to the American hold on the genre is Thomas Hylland Eriksen's *Tyranny of the*

Moment. Written from a Norwegian slant, it argues that despite the availability of technologies like email, broadband, the internet and mobile phones to speed up our lives, we are still short of time. Our email inboxes overflow and our World Wide Web searches put us on to a mind-blowing number of leads. The result? Data smog. Despite the increased access to information, our attention is now divided across competing informational demands. In our most vulnerable moments we are at risk of both distraction and delayed productivity. But at our best we surf, browse, search and make snap decisions.

That we make such a public display of our relationship with time reveals just how preoccupied we are with creating an intimacy with it. But poke about below the surface and there are signs that this intimacy is actually rather strained. For example, take how much spare time we think we have. '*Spare* time? Get real!' you may screech. Research duo John Robinson from the University of Maryland and Geoffrey Godbey from Pennsylvania State University gathered thousands of on-the-spot estimates of the size of weekly spare-time 'budgets' from Americans of all ages, including those in paid work as well as those not. The average stood at eighteen hours per week – enough time to watch more than thirty-six episodes of *The Simpsons*, fly from New York to Singapore, watch nine games of soccer, cure a light common cold or take your family on an extended overnight hike. Admittedly waiting for a cold to disappear might be somewhat less appealing than watching *The Simpsons*, so be efficient – watch *The Simpsons* while you decongest. Actually, eighteen hours was only half the story. When Robinson and Godbey asked respondents to keep a diary of their activities – showering, work, travel, childcare, cooking, walking the dog, watching TV and anything else respondents were prepared to own up to – this had the effect of more than *doubling* their perceived spare-time estimate. With as many as forty hours to play with, this allows enough cartoon viewing time to delude yourself that you *are* Homer Simpson. But relying on the average ensures, of course, that we ignore the people who get by with only a limited spare-time budget.

Given the domineering presence of the clock in North American and other westernised societies, the finding that people are so out of touch with the size of their spare-time allocation seems counter-intuitive. Of significance is that Robinson and Godbey also discovered a similar study in Japan where respondents estimated they had nineteen hours leisure time compared with an actual figure of thirty-nine. With the implication of these studies being that both these clock-conscious societies are out of touch with the way they spend their time, a mini-investigation beckons. First, let's dissect the on-the-spot task. Put someone on the spot and ask how much spare time they have and it is unlikely they will have a precise figure to hand. Pausing to come up with a figure requires them to pool assorted-sized chunks of time unevenly distributed across the preceding week, a bit like working behind a bar and suddenly being asked to add up from memory the cost of all the drinks you served last night. Despite the mental gymnastics required, surely this alone would not have caused people to cheat themselves out of twenty hours' leisure time? One obvious explanation for the discrepancy between the on-the-spot poll and the diary method could simply be that the former offers the perfect opportunity to give responses that conform more to personal ideals rather than reality. On being asked for their on-the-spot estimate, the 'I'm busy, I'm important' ego inflates – and down goes the leisure allocation. Another explanation for the discrepancy could be the interference that comes when mentally weighing up how long activities last. Depending on people's subjective experience, the estimate contracts or expands. After all, as the cliché tells us, time flies when you're having fun. Alternatively, respondents may pigeon-hole some activities – especially watching TV – into a non-leisure rather than a leisure category. In full-scale surveys sociologists do all their own classification of activities into groups, which means this risk of ambiguous categories is reduced.

Taken together, these various factors could help explain why Robinson and Godbey found that respondents understated their leisure time, overstated their working hours and overstated time spent on housework. In one survey they even found that once respondents' weekly time-use estimates were pooled together across all time-use categories, the total exceeded the maximum 168 hours available in any week by seventy-two hours, effectively making it a ten-day week. As it happens, a ten-day week would hardly be a novelty. Back in 1793, in an effort to eradicate religious connotations and move towards a decimal system, the leaders of the French Revolution replaced the seven-day cycle with a ten-day one. Each month consisted of three weeks called *décades*, with the days named arithmetically – *primidi, duodi, tridi, quartidi, quintid, sextid, septidi, octidi, nonidi* and *décadi*. As this arrangement meant workers had one day off in ten instead of one in seven, the calendar was not exactly a hit and Napoleon replaced it with the Gregorian one in 1805.

Despite their general reputation for being highly time-focused, US and Japanese respondents were not in touch with – or at least not able to admit to – how they used their time. Despite gaining an extra five hours spare time a week since the 1960s, Robinson and Godbey found a continual increase in the number of US citizens reporting feeling 'always' or 'sometimes' rushed, from 74 per cent in 1965 to 87 per cent in 1995. And, crucially, they were *twice* as likely to say that they had less rather than more free time. However, the Japanese and the Americans parted company when it came to their experience of feeling rushed. Whereas 35 per cent of US citizens feel 'always rushed to do things', only 17 per cent of Japanese agree.

So, sit Japan and the USA on the couch, and to different degrees a strained intimacy with time reveals itself. This is not to say that other countries might not also show a similarly impoverished relationship with time – after all, as I showed earlier, ever since the Industrial Revolution various Europeans have been vocal about living in a hurried era. But because to date there have been no country-by-country comparisons charting how much spare time people have (or believe they have) against their reports of feeling rushed, this makes it difficult to know whether Japan and America are the odd ones out. For all we know, Japan and the US may be in good company alongside a host of other countries who also have similarly strained relationships with time. Aside from this, there's another factor to keep in mind. Even if more comparisons were available on a country-by-country basis, the problem is that they would likely be biased. The trouble is that sociologists living in highly clock-aware countries often gather time-diary data for westernised societies, followed by what can only be described as a festival of number crunching. A single figure emerges for each category of time use, which is usually partitioned into sleep, meals, personal care, household work, leisure, travel and paid work. These clock-based accounts mean that judgements about people's relationships with time rest solely upon trends expressed in terms of hours and minutes. Opportunities to pick up on alternative ways of relating to time often get neglected. It is only by taking a step back and looking at the range of ways it is possible to relate to it, that a more rounded picture is possible. Get a mirror on the world's temporal landscapes, and it is easier to take stock of our own culture's relationships with time.

But what about how people and cultures vary in their relationships with time? Some research methods are clock dependent, like looking at the amount of hours that people work. Others are less clock reliant, and focus on the pace of life in different cities worldwide, the reality of the balance between leisure and work including the chance to take vacations, the nature of competing demands on children, and different cultural attitudes to time. Such methods include a focus on how the influences of consumerism, technology and globalisation are collectively taking us beyond the old nine to five.

A glimpse into the world's temporal landscapes: ways to live with time

Earlier, I traced discussion of society's time pressures back as far as the mid-1800s. But what factors led to this? One way to track developments is to visit mid-eighteenth-century Britain, where there was very little in the way of communication between cities, towns and villages. Each settlement lived entirely according to its own time zone, aided by the church bell or other timepiece set according to solar time. But in 1784 the British mail-coach service started to run according to particular schedules, and time coordination between different post offices en route became possible, as each mail coach carried some kind of time device set according to Greenwich Mean Time. This was the first attempt in history to bring previously temporally segregated communities together. The 1820s saw the arrival of the first passenger train and the subsequent expansion of the British rail network, which created the need for timetabling. Because mass timetabling depended on punctuality, this became the default relationship with time. An emphasis on punctuality was not only a characteristic of rail transport; with the widespread appearance of clocks in factories came the capacity to pay by the hour rather than piecework. Together, these developments helped to transform British people's relationship with time, and they became both coordinated within and across towns and cities as well as being clock-focused in their work hours. Parallel developments in the United States initially proved more difficult because of the sheer number of degrees of longitude involved in the standardisation of time, and even now the change to daylight saving time occurs on different weekends within the same state (Indiana).

Around 1850, wristwatches came on the market, and once they became affordable for the majority, people's intimacy with time moved to a new level. People experienced a new physical relationship with timekeeping. Instead of being separated from the body, a watch was worn next to the skin, like underwear. The extent of this attachment is revealed by the 'absent watch phenomenon'. Say you send your watch for repair or misplace it and do not have a stand-in to hand. Later on in the day you automatically look at your wrist to discover only skin. You find yourself looking at the same bare patch on your wrist again and again. As the day progresses with no sign of your obsessive checking behaviour slackening, your dependency on clock time becomes clear. If your life revolves around hyper watch-watching as you check your progress against a schedule, and if in your view time is not to be lost, saved or wasted, but managed, then you inhabit a clock time culture and are likely to be living in the US, east Asia, western or northern Europe.

Clock time cultures

In clock time cultures time is viewed as a scarce resource and a high value is placed on carving it up into activities that run back to back. In organisations with a strong clock time ethos, the prime driver is working to deadlines. Characteristic of this approach is what Edward Hall described in his 1959 book, *The Silent Language*, as monochronic time (M-time). According to M-time, tasks are typically handled in an ordered sequence, one at a time, with fairly rigid slots for appointments. Attention tends to get devoted to one conversation at a time, rather than jumping back and forth between several different people at once. Relationships with other colleagues are lower in priority than the task in hand.

In clock time cultures punctuality at formal meetings is the norm. The unspoken rule is that people make an apology if they are delayed by anything over five minutes. This may also apply for a shorter delay where there is a power difference which is perceived as important. During work time off-task talk is less welcome – about 80 per cent of the talk is about the task in hand and social chat is something to be snatched by the water cooler. Team building is concentrated into a couple of company away-days or scheduled fun at the

14

seasonal office party. It goes without saying that training opportunities in time management are abundant.

Proverbs give us an opportunity to glimpse wider cultural norms. As they often focus on our relationship with time, they serve as a highly convenient way of comparing attitudes to time across different cultures. Henry Davidoff's in-depth analysis of proverbs from twenty-five languages led to the identification of five ways time gets construed.

- Time is precious.
- Time inevitably conquers.
- Time is helpful.
- Time should be wisely allocated.
- Time can be conquered.

Proverbs with their origins in highly industrialised countries tended to be economically driven ('Don't put off until tomorrow what you can do today') and largely fell into the 'Time is precious' and 'Time inevitably conquers' categories – for example, 'Time and tide wait for no man'. This contrasts with the Mediterranean island of Sicily, where half fell under the 'Time should be wisely allocated' category – for example, 'Those who run stumble'. Sicilian proverbs centre on the importance of not rushing as well as framing time independently of the achievement of events. It is this particular approach to time's passage – the importance of events – which is the focus of the next section.

Event time

Event time tends to predominate in Latin America, the Mediterranean, the Middle East and some African countries. In these places the emphasis is on letting events themselves, rather than the clock, drive activity. People are treated as more important than appointments and schedules so the natural flow of events runs its course and the achievement of results is not stressed. Shops and restaurants close when the last customer leaves. Changes in activity occurs less as a result of the passage of hours and minutes and more as a result of an event reaching its natural end.

People immersed in event culture drop by one another's houses without arrangements being made in advance, particularly in less urban areas. In his book *Geography of Time* Robert Levine shows how in Burundi people's arrangements are not tied to a specific time. In many villages people do not wear watches; rather than organising their lives around specific times, they take a more activity-centred approach and meeting up is negotiated by arranging a loose event time by saying something like, 'I'll meet you when the cows go out for grazing.' As the actual time of grazing will vary according to conditions day to day, this might mean waiting around for an hour or so.

In another anecdote Levine talks about his visiting professorship to Brazil, where he was puzzled that only a few students turned up for the start of his first lecture. Over the course of the 'scheduled' two hours, the students walked in, smiled hello, sat down and carried on settling in apparently as normal. No one tried to creep in or saw the need to offer an apology; they just came to the lecture when they were good and ready. Punctuality was of no concern; instead the overriding ethos was time's flexibility – also known as 'rubber' time.

Event time culture is characterised by a tendency to polychronic time (P-time), which avoids having a predetermined schedule. Instead, the emphasis is on switching between activities in no strict order. In the workplace the priority is social relationships, with a 50–50 split between getting on with the task in hand and off-task talk. It is this social cohesion which is valued for facilitating team working and developing new collaborations. Latin American and Mediterranean styles rely more on mixing business with social time, and in south Asia, especially India, where arranged marriage is common, socialising and business are tied together: looking for a bride or bridegroom is also a way of seeking to expand business.

Timeless time cultures

Timeless cultures are common where Buddhism or Hinduism are practised. There is less attachment to fast-forwarding to what's going to happen next, or rewinding to what's been happening. Units of time are defined more by events than the clock. The emphasis is away from physical explanations of time and more towards the *experience* of time and individual awareness. A hallmark of this approach to time is the focus on the moment. For instance, every effort is made to complete conversations with co-workers; the default mode is to focus on the here and now. The current topic is not discarded in favour of the promise of jumping to fresh conversations with someone else.

A timeless approach can become the default relationship with time for people diagnosed with terminal illness. A way of coping is to adjust their vision of life to one of living from one moment to the next. This idea is revisited in Chapter 4, where I talk about our relationship with time in terms of our experiences.

Harmonic time cultures

In China and other places practising Taoism and Confucianism, the vision of time is one which promotes synchronicity between people and nature. The mindset is typically oriented to the future – towards building foundations for the long term. For instance, one common ritual is to plant slow-growing trees so that grandchildren may have the chance to appreciate them.

Merging time cultures

So far, people's attitudes to time have been broadly compartmentalised according to the part of the world they inhabit. But now that we live in the digital age, of course, things are not quite so clear cut. Technology propels us into new kinds of relationships with time and allows us to multitask, so the boundary between M-time and P-time becomes blurred. People living in traditionally M-time cultures can now be observed doing more than one thing – yapping on the mobile while simultaneously zapping spam emails and wolfing down a sandwich. Similarly, the introduction of technology can transform the timing of work practices from P-time to M-time. Stephen Barley at Stanford University looked at the way technicians and radiologists related to time in hospitals both before and after the introduction of computerised topography (CT) – imaging technology which uses a computer to organise the information from multiple X-ray views to construct a cross-sectional image of areas inside the body. Before its introduction, the technicians and radiographers stuck to M-time and P-time modes respectively. Typically, in their P-time mode, radiographers would switch between activities – from having consultations with physicians to reading films or talking on the phone. Their activities occurred in unpredictable sequences and involved working along multiple lines of simultaneous action, which often led to frustration for the technicians, who when they sought out radiographers for the next scheduled patient, found them otherwise engaged. Once CT became available, a new pattern emerged. The technology effectively 'anchored' the radiographers, so reducing the rate of switching between tasks and lengthening task duration. In this new M-time mode it became easier for both radiographers and technicians to work out the order of activities, the amount of time needed for each and to gain advance warning of the recurrence of particular duties.

Despite the merging of M-time and P-time cultures, it turns out that the distinction has not been completely lost. Differences between M-time and P-time cultures can still be detected in responses to download times on e-commerce sites. When Gregory Rose and co-workers from Washington State University compared people from countries traditionally associated with M-time cultures (the US and Finland) with ones from P-time cultures (Egypt and Peru) they found that P-time cultures had a more positive attitude to delays in web download time than those from M-time cultures. These findings may persuade e-commerce site directors that for websites with dedicated language or country web pages it is feasible to include rich content, even if consumers are on narrowband.

Identifying different cultural attitudes to waiting time through the medium of computers is relatively uncomplicated because consumer reactions are easily monitored online. But when it comes to finding ways to investigate our pace of living, some creative solutions are called for.

A glimpse into the world's temporal landscapes: who lives the fastest?

The difference between the pace of life in the city and that in towns and rural locations is often brought to a head when locals meet outsiders. Last spring, during a visit to the Valencian town of Oliva, as we drove our car around the winding backstreets, we frequently came to a halt as the driver of a car in front stopped for a lengthy natter with local passers-by. Similarly, on Mull and other Scottish isles the locals chat through their vehicle windows to pass the time of day with other islanders but get interrupted by time-pressured tourists stuck behind them, who get fed up waiting and beep their horns. Compare these incidents to the pace of life in the world's cities. In London drivers thump their horns and rev up their engines whenever a driver fails to set off at the earliest sign of the all clear. In Rome and Paris drivers play pedestrian roulette. The traffic reluctantly creates gaps – only slowing at the last moment – and not necessarily even at designated pedestrian crossings. But this is not universal in industrialised society; in Tokyo both traffic and pedestrians meticulously observe the traffic lights. Drivers slow down promptly in conformity with the lights and pedestrians wait patiently for the all clear. Though this may be attributable to cultural attitudes toward authority, nevertheless, even in time-pressurised Japan, city dwellers are prepared to wait.

Given that a feature cities share is that they are naturally busy centres, surely then we could expect them to have a similar pace of life? The reason it is worth noting how much they vary is that higher time urgency is thought to be linked with greater stress levels and coronary heart disease. According to Robert Levine from California State University, cities with a faster pace of life are therefore more likely to have people showing what is called – not without controversy – type A behaviour. Type As strive to achieve goals as quickly as possible, talk faster, have a strong competitive streak, have a short fuse and finish off other people's sentences for them. Levine's argument links a faster pace of life with greater stress, which in turn is often linked with unhealthy behaviours like smoking that go hand in hand with a higher rate of coronary heart disease.

In investigating the plausibility of this theory, Robert Levine compared the pace of life in thirty-one cities worldwide using three indicators. First, to get a sense of the importance attached to the clock in daily life, Levine checked the accuracy of city clocks compared to local time as measured by the international standard given by the telephone company. Next, he homed in on the walking speed of locals by getting observers to hang around during office hours lying in wait for unsuspecting pedestrians. Any solitary adult who was not obviously window-shopping became a target for timing as they traversed sixty feet. To ensure that any variation in weather conditions had only minimal influence on readings, observers only ran the study on sunny days.

The third and final indicator of the pace of life was working speed. Levine opted to study postal clerks on the grounds that their duties and working environment are similar across countries. The task, which was timed, involved local researchers posing as customers in post offices and asking clerks to fulfil a standard request for stamps. Each researcher handed a written note requesting a stamp of a commonly used small denomination to the clerk. If you put yourself into the shoes of the postal clerk, the natural reaction to a note would be to think that you were about to be robbed, and that the task did not always go to plan is something Levine himself hints at in his book *A Geography of Time*. He reports a New York City post office counter assistant's reaction on receiving the statutory note. The assistant disbelievingly assumed a slow, loud voice and said, 'You ... mean ... to ... tell ... me ... that ... you ... want ... one ... lousy ... stamp ... and ... you're ... giving ... me ... a ... five-dollar bill?'

Combining individual country ratings across these three tasks gave Levine a league table, which is reported here in the order of the fastest first.

1. Switzerland (Berne and Zurich)
2. Ireland (Dublin)
3. Germany (Frankfurt)
4. Japan (Tokyo)
5. Italy (Rome)
6. England (London)
7. Sweden (Stockholm)
8. Austria (Vienna)
9. Netherlands (Amsterdam)
10. Hong Kong (Hong Kong)
11. France (Paris)
12. Poland (Wroclaw, Lodz, Poznan, Lublin and Warsaw)
13. Costa Rica (San Jose)
14. Taiwan (Taipei)
15. Singapore (Singapore)
16. United States (New York City)
17. Canada (Toronto)
18. South Korea (Seoul)
19. Hungary (Budapest)
20. Czech Republic (Prague)
21. Greece (Athens)
22. Kenya (Nairobi)
23. China (Guanzhou)
24. Bulgaria (Sofia)
25. Romania (Bucharest)
26. Jordan (Amman)
27. Syria (Damascus)
28. El Salvador (San Salvador)
29. Brazil (Rio de Janeiro)
30. Indonesia (Jakarta)
31. Mexico (Mexico City)

Overall, the pace of life was fastest in western Europe and Japan and, consistent to stereotype, slowest in less economically developed countries. Western European countries hogged the top of the table. Notable absentees from the top were the US and Canadian cities, which ended up as neighbours halfway down the overall list. Levine's work showed that the fastest pace of life was linked to higher coronary heart disease death rates and higher rates of smoking, though Japan proved an exception. Whereas Japan came fourth in the overall pace-of-life listing, its rate of coronary heart disease was exceptionally low and in line with that of countries like South Korea, China, El Salvador and Mexico, where the pace of life was found to be slower than in Japan. That Tokyo had such a fast pace of life and such a low rate of heart disease reveals that living fast and its associated stress do not necessarily have to play such a pivotal role in the onset of coronary heart disease. Either that, or the Japanese have found the secret of living at a fast pace with low stress.

Which cities were top of the individual categories? The place with the most accurate clocks was Switzerland, where the clocks were only out by an average of nineteen seconds, compared to El Salvador, where they were out by eleven times as much. That Switzerland headed this category comes as no surprise given its reputation for the manufacture of timepieces and when it is home to trains with excellent punctuality records. The city where people came closest to resembling Olympic power walkers was Dublin. Their average time (of just over eleven seconds over sixty feet) was some five seconds faster than the more leisurely strollers of Brazil. As for speed of work, the postal clerks in Frankfurt clearly remained unsuspicious of the researcher's note, completing the task in just over thirteen seconds, five times faster than in Mexico.

Another reason to home in on a country's pace of life is that it gives us insights into how people get on. By prioritising the needs of others, people take the time to talk and to help one another rather than being consumed by the desire for individual achievement. Such community-minded approaches – often referred to as collectivism – go hand in hand with a slower pace of life. As I show in Chapter 3, active social ties are crucial because they are thought to bolster immune system functioning, which in turn protects against cancer and ultimately impacts on another of our relationships with time: longevity. There are signs that collectivism starts young. Jo Ann Farver and co-workers from the University of Southern California showed that Indonesian children are more likely to tell stories of dolls being friendly and helpful than children from America, Germany or Sweden. Countries with the highest collectivism ratings include Indonesia, El Salvador, Syria, Jordan, Bulgaria, China and South Korea. Countries with strong individualistic stances – in particular the USA, but also Switzerland, England, the Netherlands and Canada – put less emphasis on obligations to family members, and instead reward individual achievement. Famous psychologist Stanley Milgram showed that in places with a fast pace of life social responsibility is not high on the agenda and the needs of others receive less attention. In fast-paced cities where people get on with their own business they are less likely to offer help to strangers in the street or an arm to people with a visual impairment trying to cross the road. This is entirely in line with Levine's finding that in places with a faster pace of life there were fewer signs of helping behaviour. Before despairing over the lack of compassion among the population of western Europe, it may be heartening to hear that in at least thirteen countries across Europe people on average spend ten minutes a day doing volunteer work or helping other households, which amounts to around four months of our entire adult life dedicated to helping others.

Pace of life is not only of interest to social psychologists. Alongside climate, pace of life is increasingly one of the top selling points for property agents trying to entice residents from fast-paced countries to retire or buy a second home in a slower-paced one. The number of British households owning a second home in places like Spain, France and other Mediterranean countries has now jumped to 6 per cent. Some 175,000 Britons are living in Spain, with many citing the more relaxed pace of life as being a key reason for moving. Projections by the Centre for Future Studies forecast that by 2020 around a tenth of our current population will move abroad. Perhaps incomers into slower-paced countries partially adapt to the local pace, but their arrival in large numbers may also speed it up somewhat.

The popularity of the 'slow food' movement, which now counts over 80,000 members in more than a hundred countries, is another sign that people are questioning their relationships with time. It was founded by Carlo Petrini and a small group of leftists in 1986 and its goal is to keep regional cooking and gastronomic culture alive – 'to protect pleasure of the table from the homogenisation of modern fast food and life'. The idea is to conserve agricultural biodiversity and protect traditional foods from extinction. Early fruits of the movement included the Italian wine guide *Gambero Rosso* and the opening of a restaurant, Osteria del Boccondivino, in Bra, Piedmont, which served local food at modest prices. A few years later its protests against plans for a McDonald's in Rome's Piazza Navona gained the movement notoriety. One of the latest projects is the Ark of Taste, which raises awareness of local products that are economically sound and commercially lucrative but threatened by the standardisation of industry and legislation regarding large-scale distribution. Endangered products include those from the Italian Valchiavenna goat, the American Navajo-Churro sheep, the last indigenous Irish cattle breed, the Kerry, and Greek fava beans grown only on the island of Santorini. The movement's website includes a weekly recipe feature. At the time of writing this included ideas for marinated sardines, tagliatelle stockfish, fish stew and almond fritters.

A glimpse into the world's temporal landscapes: who works the longest hours?

Now it's time to return to the currency of hours and minutes, comparing the extent to which work fills our time across the world. Some 605 hours separates the annual paid working time of the workers of the world's least and most hard-working cities, which over the course of an average adult working life mounts up to a whopping thirteen years. When this figure is put into the context of life expectancy in countries with mid to high mortality rates, then it becomes clear that a great deal of some people's already relatively short lifespans are swallowed up by work.

People work the longest hours in cities around Asia, South America and east Africa, with employees in Abu Dhabi clocking up an average of forty-two hours a week. However, this figure represents an underestimate because it assumes no leave is taken. In contrast, those with the best deal are workers in the seven European cities which claim all the lower slots on Table 1, of which London is the most hard-working and Paris the least. That Parisians work fewer hours probably reflects the passing of the Aubry laws, which saw the introduction of the thirty-five-hour week in 1998. This succeeded in practice because the government ensured a high rate for overtime, but now there are question marks over the future of the thirty-five-hour week because of concerns about the perceived threat to the country's economic performance.

Table 1: Average annual work hours in thirty cities

City (average number of hours worked per year)

Abu Dhabi (2192), Bogota (2182), Hong Kong (2181), Taipei (2176), Manila (2164), Nairobi (2164), Mexico City (2150), Panama (2121), Bombay (2097), Bangkok (2092), Seoul (2073), Jakarta (2065), Buenos Aires (2005), Shanghai (1983), Los Angeles (1939), Johannesburg (1929), Rio de Janeiro (1912), New York (1882), Tokyo (1864), London (1833), Moscow (1824), Montreal (1814), Sydney (1749), Milan (1732), Madrid (1718), Copenhagen (1687), Amsterdam (1686), Athens (1686), Berlin (1666), Paris (1587)

Adapted from UBS Prices and Earnings Survey

Although the table gives us a snapshot of world diversity, working conditions for specific groups of people remain hidden. For example, according to recruitment consultancy Adecco, British managers and senior executives work an unpaid extra fourteen hours a week. But let's shift our view from this high-powered world to take a peek at conditions in the world's sweatshops, where people can be found stitching our designer clothes and trainers from 7 a.m. to midnight – or worse, doing twenty-four-hour shifts – for as little as $0.14 per hour. Possibly the most extreme conditions are found in China, where mass migration from rural poverty has resulted in millions moving to the city to find work. Migrants are prevented from gaining permits to stay in the city by the state system of resident registration and end up crammed into dormitories with as many as twenty other workers, earning barely enough to cover their food and rent let alone pay for the administration of permit applications – or cover the costs of sending their children to school. The daily reality for migrant and sweatshop workers is alternating between sleep and work, with no time for leisure activities, not achieving a work–life balance.

So far I have kept the picture deliberately general and avoided comparisons by gender. The bold opening statement of the European Commission's report on time use in Europe included the contention that women and men spend time in similar ways. But let's take a closer look. On average, the employed

European woman spends twenty-five more minutes per day juggling paid work, household and family care than does the average man, with more household work being carried out by women in every single one of the thirteen European countries surveyed. By the end of the week the cumulative difference is two hours fifty-five minutes, and by the end of forty-seven years of adult working life this amounts to an entire ten months.

Although these figures show that women spend more time in these core activities, what they fail to capture are the other relationships with time – the multitasking, the managing and the scheduling. In the UK one constant scheduling headache is school finishing times, with parents – usually women – trying to coordinate school runs with their work schedule for the mid-afternoon finish – between 3 and 4 p.m. Compare this with countries in Europe like France, where the school day typically runs from 8 a.m. to 5 p.m. and therefore fits in with office hours. However, the British government now plans 'extended schools', which by 2010 will make available to under-fourteens breakfast and after-school clubs between 8 a.m. and 6 p.m. The idea is not only designed to offer children the chance to have fun and learn new skills, but is also supposed to enable parents to juggle home and work responsibilities better.

The reality of the work–life balance

How do different countries fare in the balance between life and work? First we must consider annual leave allocations worldwide. The easiest way to do this is to compare conditions for workers employed in the same line of work. Table 2 shows the situation across the world in manufacturing. Workers in European countries are not only allowed the most holiday but employers are positively pro-vacation. In the Netherlands some employers run schemes which enable employees to put money aside out of their salaries to save up for holidays. Elsewhere in Europe factories and plants either partially or completely close down and workers take the entire month off. In France there is even a name for employees who take July off – *Juilletistes*, whereas those taking August are *Aoûtiens*.

Table 2: Number of days annual holiday for manufacturing workers

Area	Annual Holiday	Public holiday	Total holiday	Work hours per week
Netherlands	32	8	40	39
North Europe	26	8	34	39
South Europe	22	12	34	41
East Europe	18	8	26	40
Asia	14	12	26	44.6
Africa	15	9	24	42
Latin America	15	9	24	44
USA	12	11	23	40
Japan	10	13	23	42

Adapted from Richards (1999).

Although Japanese workers are authorised to take twenty-three days' leave, many take as few as eight, which restricts travelling abroad for relaxation, yet surveys show that over a third of Japanese long for an overseas holiday. Moreover, a high proportion of vacation time comprises single-day public holidays, which means less flexibility for the workers about when they take their holidays. The impact of this pattern on the biology of rest is considered in the next chapter, but one problem to note here is that the Japanese work ethic is so strong that in 1995 there were sixty-three official cases of *karoshi* – death through overwork. One reason workers did not take holidays was that they feared their jobs would not be there when they returned, although the Japanese government has taken steps to encourage holidays by saying 'to take a vacation is proof of your competence'. In the US vacation allowances are relatively low, but there are signs that the balance between paid work and life is changing. Workers are opting out of long hours with nearly a fifth of the US adult population choosing to adapt their working hours and reduce their salaries.

Leaving aside vacations and turning to leisure time, Figure 1 shows that Europeans have never had it so good. In the 1960s the UK scraped by with only 1.26 times as much leisure time as paid work. But since then there has been a sharp increase in the time devoted to leisure relative to the total

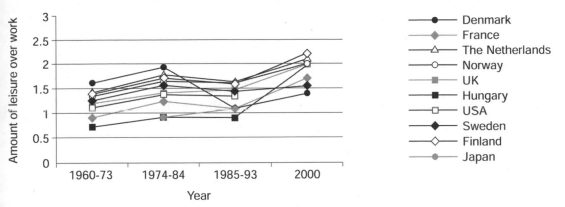

Figure 1 Increase in leisure time over time spent in paid work.

number of paid hours worked. The amount of leisure hours in Europe is now around double the amount of time spent in paid work.

During the last thirty years, Hungary, France and Japan have all had periods where time spent in paid work exceeded leisure time. This has improved, but at the last count Japan was still bottom of the league. Some 85 per cent of US workers and 92 per cent of French workers work a five-day week, while only 27 per cent of the Japanese labour force work as *little* as this.

If you're looking to relocate within Europe, which country fares best in terms of time spent on leisure as compared to paid work? Belgium now enjoys 2.6 times as much leisure as paid work, followed by Romania, Finland, Estonia, the Netherlands and Norway. Those countries enjoying more leisure than paid work but not yet twice as much include Hungary, the UK, Slovenia, France, Sweden, Denmark and Portugal. Despite being leisure-rich, the possibility remains that Europeans work and play long hours at the expense of sleep. Take Norway, a country where combined leisure and paid work hours are the highest in Europe; the Norwegian government reports that sick leave rose by 12 per cent between 1999 and 2000, and it is tempting to think that a link exists between this and lack of sleep. Compare the figures for France, where paid work and leisure combined is the lowest in Europe (which tallies with the finding for Paris in Table 1). That the French are found to sleep more than any other Europeans – over nine hours per night – shows that what they lose in leisure and paid work, they appear to make up in sleep.

The trouble with concentrating exclusively on the size of leisure allocations is that this gives a misleading impression of our relationships with time.

After all, for most of us leisure is not taken 'at leisure'. Instead, the default relationship is needing to 'make' time – usually in isolated short bursts rather than extended, unbroken chunks. Leisure is squeezed in between blocks of time already constrained by other scheduled commitments. Dale Southerton from the University of Manchester talks about our attempts to deal with time using the concepts of 'hot spots' and 'cold spots'. Things like work, chores and appointments get compressed into scheduled hot spots because they are dependent on other people, networks and services being available at particular times. This scheduling of hot spots frees up cold spots during which family and friends can preside. Although scheduling is a highly adaptive coping strategy to deal with concerns about time, it creates a high mental load, giving rise to feelings of rushing. For example, in clock cultures, where scheduling dominates, there is constant monitoring of the extent to which any activity is on track to be completed within the available time slot, before the next appointment. Not only that, but as there will be rapid switching between different types of activity, there will be increased memory load due to the need to carry over an assortment of things to remember until the next activity.

Southerton notes that even scheduled cold spots can take on the status of hot spots because of the need to travel to a variety of leisure-related locations within the course of one day.

> We started off early on Saturday and I did the cleaning on Friday night, just to free Saturday up...So we've done breakfast and the bathing and we went swimming, and took a picnic so we had that, and then went on an adventure trail at Bowood. And because we had to get there by twelve to make it worthwhile we had to leave swimming with just enough time to spare. Then we had to find somewhere for the picnic and it just went on like that. God I needed a day off after that day off!

The concept of the work–life balance is not the most appropriate one to adopt in the context of the 'scheduled society'. This latter approach to time treats it as something that can be neatly partitioned into slots for different purposes. With the reality being rather different, there is an urgent need for government and policymakers to switch from this traditional discourse to using words that capture the daily challenges people face in moving through hot and cold spots of time while trying to coincide with others.

Of course it is not only adults who learn to juggle competing demands; the vast majority of the world's children do so too from an early age. In event time cultures, however, this juggling may be less clock driven, and determined more by the amount of available daylight hours.

A glimpse into the world's temporal landscapes: competing demands on children

Between a third and a half of British schoolchildren have part-time work, with newspaper delivery being the most common job. As a teenager I had a paper round, stuffing papers through letter-boxes (often of matchbox-sized proportions). The positive aspects of the work were the sense of completion after emptying a bag containing around sixty papers and the chance to keep fit. The downsides were lugging around the full bag of papers and on weekdays the challenge of making sure the round was finished before the start of school. And of course the pay then – as it is today – was a pittance. But compare this situation to that of the young newspaper vendors of Ethiopia, who not only have the pressure of juggling school and work but also need to make sufficient sales to both support their family and fund their schooling. They sell papers in the morning when there is most demand, and go to school in the afternoon. A project by Martin Woodhead at the Open University found similar long work hours and competing demands on children's time in Bangladesh, the Philippines, Guatemala, El Salvador and Nicaragua:

> Ten-year-old Russell works as a porter in a Dhaka market. He's already looking for business by 8 a.m. each morning, and works through until 7 or 8 each evening, with just a short break to go home for lunch. He grew up in rural Bangladesh, but when he was 7 his family fell on hard times, and were forced to look for work in the city. He got as far as Grade 3 in the village school. But now his father is sick and unable to work much, so school is out of the question for Russell: 'If I stop working today and go to school, who will feed my family? It would mean earning and eating less because I would have to divide my time.'

In Bangladesh children start work as brick chippers when they are old enough to hold a hammer. From the age of ten they work for around ten hours a day, starting at 7 or 8 a.m., working two- or three-hour periods with gaps for domestic chores – collecting water, washing and so on – until finally settling down around 11 p.m. 'Our mothers bought us small hammers and we used to break the bricks for fun. It was like playing. Then gradually when we learnt how to use the hammer our mothers asked us to start working.'

In the Philippines Woodhead found that the majority of children were doing five full days of schooling, two hours of fishing a day and domestic chores. During the weekends and school holidays the entire day was normally spent fishing, gathering shellfish or dressing chicken. Children in some rural parts of Guatemala get up at around 5 a.m. to do chores, then start work, being paid piecework to make fireworks. Typically, the job involves making a tube for the firecracker, filling it with gunpowder, fitting the wick and packing the finished product in boxes. After this, they do a morning at school followed by another shift at work. In Mayan communities child farm workers typically miss the last month of the school year as well as the first six weeks of the new year to pick cardamom and coffee for around ten hours a day, all the while trying to avoid snakes and mosquitoes.

In these childhoods there is no formal separation of work and leisure, and children contend with the pressures of work, school and supporting their families from early on. These competing demands mean that there are limited opportunities for them to call any time their own, though the work of Samantha Punch of the University of Stirling suggests that rural children seem to fare better than those living in more urban areas. When Punch lived with families in a rural part of Bolivia, she found children used their homeward journey from school to play in the river, have games of marbles or pick flowers with their brothers, sisters, cousins and friends before starting their chores.

For reasons of poverty, lack of resources, discrimination or political unrest around 20 per cent of the world's children never have any experience of schooling. The remaining 80 per cent may only spend three or fours years of their lives in education. Instead, they undertake forms of labour some of which are extremely hazardous, and can end up working more than the average adult working hours cited in Table 1. One of the worst forms of child labour is soldiering. Some 300,000 children in the world aged between five and fourteen currently carry out the same duties as adult soldiers – guard duty, patrolling, checkpoint duty, cooking, cleaning, front-line fighting and executing suspected enemies. An inquiry by Astri Halsan Høiskar from the International Peace Research Institute in Oslo found governments in Afghanistan, Algeria, Angola, Burundi, Colombia, the Democratic Republic of Congo, Ethiopia, Burma, Russia (Chechnya), Sierra Leone and Sudan all forcefully recruiting under-fifteens solely to go to war. In some cases it was not just one generation that served; the Ugandan National Resistance army included young mothers with babies strapped to their backs. Though the International Court of Justice attempts to bring government officials to trial for forcing children into battle, during wartime international law is often disregarded and the availability of lightweight weapons means that ever younger children are being conscripted.

Probably the world's largest concentration of child labourers is in Sivakasi in the Indian state of Tamil Nadu, where some 45,000 children living in villages are picked up by bus – as many as a couple of hundred per vehicle – and taken to firework factories. They work around twelve hours and do not get home until after 7 p.m. In the run-up to Diwali, the festival of lights, they work a seven-day week. The corrosive gunpowder mixture is not only an irritant to the children's hands, but contains chemicals that lead to breathing problems and blood poisoning.

Around urban centres in many developing countries children do not differentiate between home and work because they live without a roof over their heads. Some children spend between eight and ten hours a day switching between begging, caring for toddlers, drug peddling, prostituting and scrounging from garbage. Between these activities they must find edible food and a safe place to sleep, all while trying to avoid illnesses like cholera, typhoid and gastroenteritis. Street children with access to shelter typically divide their time between chores, cleaning, cooking, washing, fetching water and collecting wood. In the small amounts of time that could be described as leisure, such children chat, dance or play music with gang members, friends, brothers and sisters. Contrast this existence with childhood in westernised society, where many children's ride home from school is followed by their weekly activities and homework. If parents want to give their children opportunities they did not have themselves, there will be much to schedule in – Saturday morning dance class, gym club, Cubs, football and so on. Kitchen calendars are jam-packed with riding lessons, birthday parties, school trips, school fairs and sports tournaments. Particularly in dual-income families, children's bedrooms are leisure havens equipped with TV, DVD, PlayStation and radio. A revealing slant on this consumer-driven lifestyle comes from Glover Ferguson, chief scientist at Accenture Ltd. Ferguson believes that if we had frozen our standard of living in 1950 and applied subsequent productivity gains to reducing our work week instead of increasing our consumption we would now work two days and enjoy a five-day weekend. Instead, current consumer and work patterns in westernised societies revolve around an any-time mentality – people expect to be able to engage in the activity of their choice at any time of day or night.

Beyond the nine to five

'The mass timetable of the industrial world, of the "9 to 5" office world, and of silent Sundays, has given way to a flexi-time, flexi-place world of the new economy.' (Hardill and Green)

The latest European survey (based on fifteen countries) showed that only 24 per cent of people now work office hours, defined as between 9 a.m. and 6 p.m. Monday to Friday. The changes in shopping hours in northern Europe over the last three decades reveal why. Back in the 1970s the most popular time for shopping was Saturday morning, but in for instance Norway there has been a rise in shopping trips occurring on weekday evenings from 6 to 17 per cent. In Britain Sunday trading became legal a decade ago, and since then there have been more opportunities for shopping resulting in demands on retail workers to be available for longer shifts, distributed around the clock. In large towns and cities, the 9 a.m. to 5.30 p.m. opening hours have evolved, with retailers typically open between 10 a.m. and 8 p.m. on at least one day a week. In the early 1990s there was just the safety net of the local twenty-four-hour petrol station to pick up your forgotten milk and coffee (and even things like bird baths or other garage forecourt paraphernalia). Nowadays, late-night convenience stores are common, and many UK cities have a twenty-four-hour supermarket open non-stop Mondays to Saturdays. In inner Amsterdam too, shops are open seven days a week, with most grocery stores open eight until eight, six days a week. This pattern contrasts with many southern European cities, where 24/7 supermarkets are extremely rare.

In Rome, Milan and Athens shops stay closed from 5 p.m. on Saturday through to early Monday afternoon. The pattern of closure during the middle of the day also varies. In many cities throughout Spain, France, Italy and Greece shops close for a couple of hours at lunchtime and then reopen for a few hours in the early evening.

Technology has been instrumental in ensuring we work beyond nine to five. The widespread use of notebook PCs and mobile communications has shifted the possibilities for work into new times and spaces. Instant electronic communication allows round-the-clock working from home, car, train, airport, motorway service station, hotel lobby, coffee lounge and client office. Technology plays a key role in evolving relationships with time in the work-place. The twenty-first-century worker not only needs to be technology-literate but 'flexible' – willing to work at home, on the move or in some kind of designated remote centre. Around a decade ago commuter trains were relatively free of the carnival of assorted ring-tones and the sound of fingertips hammering laptop keyboards. Now the train carriage serves as a mobile office. In Sweden the X2000 high-speed train has business cars equipped with phone, fax and photocopying facilities. The Norwegian equivalent – the Signatur – is fitted with sockets for laptops. In the UK Virgin rail has also equipped many customer seats with sockets. The only place on the train where the boundaries between work and home are kept at a distance is the 'quiet coach'. For employees, therefore, the attractions of so-called flexible working may come at the cost of being ever available, and preferably not in the quiet carriage.

Of course, what flexibility means in practice depends on the extent to which the employer and the employee meet halfway. What flexibility buys companies is the scope to alter hours of operation to suit season, competition and demand. This brings with it some opportunity to negotiate the right to extend or reduce the length of the working week without varying wages. The ideal scenario for the employee can obviously only be defined in terms of individual preference, but typically will give scope for the employee to pursue whatever is personally meaningful to them at their particular life stage, whether this is care responsibilities, retraining for a second career or com-muting to a distant family base. In practice, this may mean the option of switching shifts between workmates, working part time, opting for a compressed working week, or altering start and finish times. Also, it often involves home working. In British workplaces employing more than five people around 20 per cent of employees work at home. Flexible work blurs the boundaries between home and work, and thus the traditional line between work and leisure time. Many professionals are required to be mindful of time allocated to personal tasks, projects and activities, to treat time as a precious resource, yet also need to be on call at all hours as this is perceived to be necessary to meet fully the expectations of their employers.

What is the impact of flexibility on health and well-being? If the employee works under conditions of company-driven rather than employee-driven flexibility, this has been shown to lead to a greater incidence of a range of problems including backaches, stomach aches, headaches, respiratory difficulties, heart disease, injuries, stress, fatigue, sleeping disorders and job dissatisfaction. In addition, work hours may be incompatible with those of family, friends or colleagues. This is particularly difficult when partners' work finishing times are unpredictable, or when colleagues' schedules fall out – as they often do with mobile and flexible patterns of work. In families where one or both parents do shift work, certain combinations may mean that, just as one is off to bed, the other is waking up to face the day. During the typical weekday, the average British couple spends just fifteen minutes a day enjoying a social life together.

Creating yet another coordination problem is the current trend of women delaying childbirth. Britain currently has the highest average age for first-time motherhood in Europe, which creates a new kind of spacing between generations. In Britain around one in ten women are caring for a parent while still having a dependant aged under eighteen in the household. The spread of duties across two different generations will intensify the challenge of trying to secure alignment across, for instance, school runs, working hours and preparing meals.

Time as a pressing cultural issue

On the face of it, out of the whole world, the people of western Europe appear to have the least to feel pressured about. After all, they work the lowest number of hours in the world, get authorised the most holiday, and have enjoyed an unprecedented increase in leisure time. Contrast this with the life of Sivaskian child labourers who work from 3 a.m. to 7 p.m., or Chinese migrant workers who sometimes slave twenty-four-hour shifts, with few chances for talk, rest or play just to keep the rest of the world in cheap fashion. Given the obvious advantages the time-rich western Europeans enjoy, why do they still feel so time-pressured? One reason may be that their pace of life is the fastest in the world. They run their lives according to the tyranny of their accurate wristwatches as they dash between scheduled back-to-back activities. That the risk of coronary heart disease is also higher in these fast-paced countries points to the impact of urgency on another relationship with time, that of longevity, the focus for a later chapter. However, that fast-paced Japan has such a low rate of coronary heart disease indicates that pace of life does not *have* to be such a critical risk factor, and later in the book I put the Japanese diet under the microscope and show that it may ameliorate the negative effects of a fast life.

Another reason why western Europeans are likely to feel time pressure is that in common with other highly industrialised societies the combination of flexi-time and back-to-back scheduling means that family, friends and colleagues become temporally segregated. In Europe the highest proportion of shift workers is now in the UK (23 per cent), followed closely by Spain (22 per cent), Italy (21 per cent), Finland (20 per cent), with the lowest level in Portugal (8 per cent). Such working hours might include anything from a permanent shift at a set time of day to one that rotates through morning, afternoon and night. This anytime attitude to work and consumption is leading to increased colonisation of the night, and in the next chapter I show how this trend has significant repercussions for another of our relationships with time, the synchrony between the light–dark and activity–rest cycles.

Chapter 2

TIME AND RHYTHMS:
THE CASE OF STAYING UP
ALL NIGHT

All living things, from plants and bacteria through to humans, are exposed to the daily change in natural light intensity that allows them to organise their rest and activity according to twenty-four-hour cycles. When electric lighting was first installed for domestic use in the 1880s, this gave people the freedom to redistribute these circadian cycles. Instead of having to restrict patterns of rest and activity to coincide with the natural switching from dark to light, there was the scope to entertain a new relationship with time. As some 20 per cent of employees now work by artificial light outside the hours 7 a.m.–7 p.m., we are witnessing the demise of nine to five and the familiar structure of the day. Let's take a tour of the world's industrialised societies to see what form the changes have taken.

Westernised societies: open all hours?

In order to stay competitive in an increasingly global market, businesses need to offer consumers flexibility, convenience and instant access to services day and night. In many urban economies around the world the extent of this often depends on which day of the week it is. The last show at the Empire cinema in London's Leicester Square on a Friday night is at half past midnight, but the following Tuesday, it is at 9.30 p.m. In Tokyo's Roppongi Hills district, Thursday through to Saturday you can view nine galleries of modern art in the Mori Tower museum until midnight. That the cinemas, observatory, outdoor theatre, restaurants and shops stay open after midnight show that Seattle isn't the only place for the sleepless.

Traditionally, emergency service staff have done the bulk of night work, but as the internet never closes, so shopping purchases are shipped out 24/7, which demands more flexible working. Now statistics from around Europe suggest that since 1995 the number of people working shifts has increased by 7 per cent and the hard core of emergency workers has been joined by employees from security firms, supermarkets, radio, television, call centres, banks, hotels, all-night cafes, production lines and so on. From corner shop to hypermarket, all compete for our custom. E-commerce sites like Lastminute.com offer on-the-spot deals, gifts and entertainment and are updated on a daily basis. Banking services allow us to access account balances by phone on twenty-four-hour automated systems. In the UK you can phone NHS Direct, staffed by nurses, at any time for medical advice. These services mean that the user or consumer no longer needs to plan shopping and service requirements in advance.

Last weekend I was on the train heading home from Edinburgh when I overheard a North American accent complaining, 'Everything shuts so early over here, it's such a pain.' Although it is true that shopping and service hours are generally more restricted in Europe than they are in North America there are some exceptions. For instance in Canada, the province of Nova Scotia continues to ban large grocery stores, liquor outlets and large retailers from Sunday trading. Compare this to downtown Toronto in Ontario, where Sunday trading and twenty-four-hour service signs are dotted all over the main streets. A free newspaper called *Toronto 24 hours* serves to strengthen the city's always available image. Jump over the border to Los Angeles, Las Vegas and New York and you'll find an even broader choice of facilities and services open 24/7. Sport and entertainment venues, fitness centres, casinos and all kinds of diners and restaurants stay open throughout the night.

Cities looking to market themselves to the world are often promoted as open all hours. The 24/7 badge means 'happening' and 'hip', which in turn attracts tourists, clubbers and people coming into the area to live, work and contribute to the local community. Property developers keen to lure buyers will advertise residential developments as 24/7, suggesting that they are situated among services open around the clock. The reality in many cities however is that only a minority of leisure venues stay open all hours. More reliable evidence of the existence of a 24/7 culture is the redistribution of people's daily routines. The greater the distribution of routines and activities across the day, the more societies have evolved into a 24/7 existence. Leon Kreitzman, author of *The 24 Hour Society*, reports that call traffic on British Telecom at 4.30 a.m. has increased by 400 per cent since 1989. Andrew Harvey and co-workers at Saint Mary's University in Canada looked into the distribution of sleep patterns across six countries (Canada, Norway, Sweden, the Netherlands, Italy and Austria) and found that in Canada 11 per cent of the population were awake at 3 a.m., compared to the other countries where only as many as 2–3 per cent were awake at this time. In line with what my North American passenger implied, there is currently more twenty-four-hour living in North America than in Europe, but, given the creeping Americanisation of Europe, we may be heading the same way.

If businesses increase their opening hours to outmanoeuvre their competitors, employees will be offered opportunities to do night work. But giving customers flexibility comes at the cost of a fatigued workforce frequently suffering sleep-related disorders and work-related stress. Workers may be out of sync with family and friends, and if they have changing shift patterns it is even more difficult for them to coordinate and plan activities on a week-by-week basis.

The perils of night working

The *Economist* has reported that the growth of the twenty-four-hour economy has achieved a reduction in unemployment, at least in North America. But at what cost? Flies and small mammals forced to be active outside their normal waking hours show a shortened life expectancy. Whether being up at night is for work or for leisure, doing so on a regular basis over a number of years leads to risks to health claimed to be equivalent to smoking twenty cigarettes a day. We know that doing night shifts on a regular basis increases the risk of heart disease in humans by as much as 40 per cent. Eating patterns alter at night because workers are unable to mirror their daytime pattern of eating three meals a day. If they did, this would mean eating the equivalents of breakfast at 8 p.m., lunch at 1 a.m. and dinner at 7 a.m. Instead, they tend to graze on a series of snacks. Indigestion and bloating are common and the risk of peptic ulcers is between two and five times higher for night workers than for those who work in the day. Mental health risks include increased susceptibility to chronic stress, major depressive disorders and general fatigue. The conclusion that women's reproductive health may also be adversely affected comes from observing more frequent irregularities in monthly cycles and more babies born prematurely. And no one yet knows the impact of prolonged altered light–dark cycles on the survival and health of the next generation.

The sleep debt resulting from the cumulative effect of night working causes a catalogue of performance-related problems due to slower reaction times. These include impaired memory, reduced motivation, fuzzy thinking leading to errors, a tendency to cut corners, loss of empathy and a worsening in performance in work involving fine hand movements. A study of US physicians on call who had less than five hours in bed per night showed that only 3.7 hours were spent asleep. The following day the physicians were so tired that on a standard scale measuring sleepiness they scored at a level requiring clinical intervention.

And it is not only night work that causes problems. Morning shifts that start before 6 a.m. have been shown to lead to a higher risk of severe sleepiness than shifts starting after 8 a.m. If the work is repetitive or monotonous this makes it worse. Train drivers who do early-morning shifts face greater difficulties when the time between stations is over fifteen minutes than when there are shorter gaps between stops.

The social costs of night work are high. In the United States marital breakdown and divorce rates increase by a factor of six when one partner works at night. Night workers forgo the chance to spend time with family or socialise with friends. However, despite this obvious lack of coordination between night working and more traditional schedules, the image of night work as unsocial is fading. Employers are now not only offering fewer financial incentives for evening and night shifts, they now expect employees to want to commit to working at least some evenings and nights.

At a time when society's patterns of sleep and wakefulness are being redistributed round the clock and away from the natural cycle of light and dark, the long-term health risks of this redistribution are beginning to become clear. Those who consistently stay up at night will experience the effects of the clash that occurs between their own biological clock and the natural light–dark cycle.

The rhythm of our biological clock

All types of plants, animals and bacteria have internal master clocks, which generate circadian rhythms that allow synchronisation with the light and dark cycles created by the earth's rotation and inform our bodies to prepare for rest or activity. These master clocks are so influential that even when all the twenty-four-hour cues are stripped away our rhythms continue to show troughs and peaks roughly according to this cycle – though as the oscillations of these biological clocks are not exactly twenty-four hours, each day they are reset by the natural light–dark cycle. Our body temperature, blood pressure, hormone levels (for example, melatonin, cortisol, prolactin), immune system responsiveness, alertness, mental performance and wakefulness all vary as a result of this cycle. Generally, our alertness, performance and metabolism peak in the late afternoon and reach a low point between four and six in the morning. But there are differences between people: 'owls' are more alert in the evening, and have a body temperature circadian rhythm running a few hours behind 'larks', who function best first thing in the day. This is not just about variation in physiological rhythms. Extreme morning types are more likely to have mishaps – ranging from bumping into others, not remembering someone's name and forgetting the location of keys – in the evening rather than at other times of day. Surprisingly, the reverse is not the case for extreme evening types. For them, slip-ups are spread more evenly across the day rather than being clustered around any one particular time point. Morning types are also more likely to have a lifestyle characterised by regular mealtimes, bedtimes, exercise and allocated slots for particular activities. As a result, their sleep is better in quality and this potentially has long-term positive effects on health and well-being.

The discovery that body temperature runs according to its own daily rhythm without cues from the natural light–dark cycle was first made in 1938 when Nathaniel Kleitman and Bruce Richardson looked at what happened to sleep and alertness levels when they spent an entire month living 400 metres underground in Mammoth Cave in Kentucky. Kleitman noticed that his performance on tasks like hand steadiness, card dealing and arithmetic was best throughout the morning, and reached a peak coinciding with his body temperature around the middle of the day, despite the fact that he had no contact with the daily cycle.

Some three decades later, Jurgen Aschoff from the Max Planck Institute for Behavioural Physiology upgraded from cave accommodation to purpose-built facilities in a former basement bomb shelter below Munich hospital, adapted to become a 'time-free' environment. He discovered that in constant darkness and in the absence of cyclical environmental cues, the bodies' rhythm nevertheless beats on. In later work by Aschoff and his colleague Colin Pittendrigh, volunteers were isolated from external time cues but allowed to sleep, eat and have the lights on whenever they wanted. Male staff who lived with the subjects for the duration of the study even shaved at odd times of day so as to reduce the availability of social clues about time. Under these conditions, the circadian rhythm ran at twenty-five hours rather than around twenty-four, but later this was found to be an overestimate because the presence of artificial lights reset the inner clock. More recent studies have shown that our inner clock actually runs according to a cycle of twenty-four hours and eleven minutes independent of the light–dark cycle, and that the patterns of body functioning – including body temperature, blood pressure, digestion, reaction time and alertness – rise and fall according to this cycle.

Being up all night

If you stay up all night, be it for work, social reasons or necessity, then you will know only too well that sleepiness intensifies most between four and six in the morning, exactly at the point when your metabolism, alertness and performance are at their lowest. At this time, it is common for workers to nod off, even when they are trying to fight sleep. Another reason for sleepiness at the end of the first night of being awake or the first night of a shift roster is that the cumulative hours awake since last sleep is already around twenty hours. Symptoms of general sleep deprivation have been likened to the effects of drinking alcohol. According to Sally Ferguson of the Centre for Sleep Research, South Australia, after this much sleep deprivation performance levels are down to the equivalent of, or worse than, those seen at a blood alcohol concentration of 0.10 per cent, which is over the UK, USA and Canadian limit. That many night workers drive home after their shift creates a high risk situation for them and for other road users. Even when traffic volume is taken into account, the risk of an accident occurring is twenty times higher at 6 a.m. than at 10 p.m. Work shifts over twenty-four hours have been shown to increase the monthly risk of a motor vehicle crash by 9.1 per cent.

After staying up all night, it is difficult to get a full night's sleep during the day because body rhythms peak in the late afternoon, interfering with sleep. Physiological functions are still working for daylight hours. Even after years of night working, the body can still be oriented to daytime, so why does it not fully adapt? The problem is that workers get a dose of light on their way home in the early morning, which resets the clock. One way to reduce the effect is to give workers twenty minutes exposure to bright light or natural light between 3 and 4 a.m. This not only has the effect of reducing sleepiness just before the circadian low – especially on the first two nights of the shift roster – but it also means that workers' daytime sleep is longer.

Where is the biological clock?

Our master clock is located deep in the brain in the hypothalamus, is called the suprachiasmatic nucleus (SCN) and comprises about 20,000 neurons. Figure 2 shows that when light hits the retina, this activates a pathway called the retinohypothalamic tract, which runs from the retina to specialised cells – melanospin cells – in the SCN. Activation of the melanospin cells encourages a reaction in various cells which act as 'pacemakers'. Although each cell is capable of pacing independently, oddly not one of them seems to take the lead. At the moment, little is known about the ways these cells 'talk' to each other. We know that light is responsible for generating and maintaining our circadian rhythms because people who have had both of their eyes surgically removed do not react to light in the same way. These individuals do not synchronise to the local light–dark cycle, which goes to show how essential the photoreceptors in the eye are in the light entrainment of the circadian rhythm.

At dusk, our pineal gland starts to secrete a large dose of the hormone

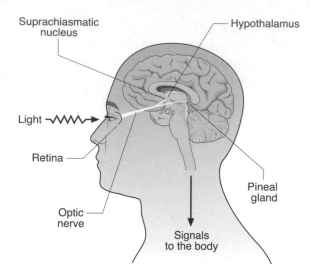

Figure 2 **Effect of light on the eye.**

melatonin, which continues during the night until there is light of sufficient intensity, duration and spectral quality to suppress production. This cyclical production of melatonin helps us to regulate sleep in relation to the duration of darkness.

As we age, the crystalline lens in our eye transmits less light – up to ten times less than in our youth – and during daylight hours the body's ability to suppress melatonin becomes less efficient. This is particularly the case when we are in indoors where surfaces are illuminated by artificial light, which is as low as 500 lux compared to 5000–100,000 lux outdoors. As this reduces the difference between the light and dark components of the cycle, and substantially reduces the amount of light entering the eye, it has most effect on older workers doing night shifts. This may disturb the sleep–wake cycle, and bring about detrimental sleep loss.

As well as having a pivotal role in circadian timing, new research shows that the SCN has a major role as the brain's internal calendar. If day length does not meet the brain's internal expectations, disruption sets in. The onset of short daylight hours in the winter months can bring depression, lethargy and feelings of demotivation – a collection of symptoms known as Seasonal Affective Disorder (SAD). In Russell Foster and Leon Kreitzman's enlightening book *Rhythms of Life* they compare the 3 per cent of people affected in the UK by SAD with the 24 per cent affected in Norway. If the reduction of light at northern latitudes has a worse impact, are there any remedies? The antidepressant effect of bright-light therapy shows positive results after around three weeks, if a light of around 6000 lux is administered in the morning for around 1.5 hours and if the affected person sleeps at the optimal point in their circadian cycle.

Genes for timekeeping

Around a dozen genes for timekeeping have now been discovered. When such genes are removed or altered, small animals typically show altered patterns of behaviour that indicate rhythms are running at the wrong speed. What are the mechanisms behind this? Each time the master clock completes a new twenty-four-hour cycle, various genes work in interlocking feedback loops to communicate the timings of the master clock to the rest of the body. Steve Kay and John Hogenesch from the Scripps Research Institute have counted as many as fifty genes that cycle in synchrony throughout the day across the liver, kidney, aorta, SCN and so on. One gene is called Rora, and produces a type of protein that binds to DNA and can effectively flick gene expression on and off. So the expression of one gene will flick on the expression of a second gene, which turns off the first gene, which turns off the second gene, which flicks the first gene back and so on, to keep the body synchronised to a twenty-four-hour day. This applies to all the clock genes – not all of which are yet known – which will overlap in different feedback loops. As a result of staying up all night, travelling across multiple time zones, or a change in seasons taking place, the clock will shift, and as different body tissues respond to the clock in their own way, they all reset their clocks independently of one another. The clock in the heart has constant surveillance of the master clock with a kind of a 'rapid response unit' approach to any detected change in circadian rhythms. In contrast, the liver has a slow response, taking several days to catch up with changes in the master clock, which makes it one culprit for the problems shift workers have adjusting to being awake at night. If you happen to be someone who puts your liver through a heavy workload due to excessive alcohol consumption, then you will have already experienced disturbed sleep. Drinking patterns tend to be heavier on days off, so let's take a more general look at the nature of the sleep–wake cycle during these times.

What happens to the sleep–wake cycle on days off?

Before the decline of the nine to five, a typical full-time employee would work Monday to Friday followed by two days' rest. On these days, people stay up late and have lie-ins. Now that there has been a shift to more people working on Saturdays and Sundays, people's lie-in days can end up being distributed throughout the week. The timing of this interruption to routine is critical because it is on days off that workers show different sleep patterns. Of course, social life goes hand in hand with days off, particularly on Friday and Saturday nights. However, the delay in sleep cannot simply be put down to late-night movies or parties. On days off, the delay in the sleep–wake cycle (falling asleep and waking later) can be anything from around thirty up to ninety minutes. Why? While the extension of sleep is due to the recovery required from sleep deprivation incurred earlier in the working week, the reason for the delayed sleep–wake cycle is attributable to the time schedules of work and school imposed in the week. As the human circadian rhythm runs just over twenty-four hours, this means that there is a potential lack of synchrony between the biological rhythm and the imposed structure of the work hours or care commitments.

The sleep–wake cycle on vacation: cultural differences

Earlier, when I compared the amount of vacation allowance between workers from different countries, one feature was that countries differed in their style of time off. In Japan a high proportion of days off are single ones. Holidays are taken in piecemeal fashion in comparison to western Europe, especially in France where it is routine for an entire month to be taken off and many families migrate to the Atlantic and Mediterranean coasts or to a mountain resort. During longer holidays the series of days off allows the timing of sleep to take its natural course – rather than committing to early-morning activities – and allows synchronisation of the activity–rest cycle with the local light–dark cycle. In contrast, those UK tourists wanting a guarantee of good weather and heading for the Caribbean, Florida, Thailand or New Zealand, expose the body to the effects of crossing several time zones in quick succession. Although on holiday, where there is normally plenty of scope to recover from sleep deprivation, being de-synchronised with the local light–dark cycle takes its toll on the body.

Travel and jetlag

On 17 December 1903 Wilbur and Orville Wright launched powered flight; a century later the aviation industry transports over 1.25 billion passengers annually. This increase in mobility not only reduces people's dependence on rail and shipping infrastructures but allows people to switch rapidly between time zones. On journeys involving travel across more than two time zones symptoms arising from the internal clock being out of step with the local light–dark cycle include disturbed night sleep, daytime fatigue, disorientation, a change in bowel habits, decreased enjoyment of food, loss of concentration and increased irritability. It generally takes one day for each time zone traversed to adjust to the new light–dark cycle. On the first day all the body's rhythms are equally out of phase but by the second and third the rhythms of different organs and tissues are all out of step with each other because they run to different clocks. The liver takes as long as two weeks to catch up. As alcohol is known to disrupt the circadian cycle, the combination of jetlag and boozy nights out abroad may push feeling below par to new levels.

To mimic the body's melatonin high when we are out of synchrony with the local light–dark cycle, melatonin is available over the counter (in the USA and Poland), as a remedy for reducing symptoms of jetlag, though the health risks associated with longer-term use remain unknown. An alternative is to plan a schedule of when to expose yourself to bright light and when to avoid it. For instance, if you normally wake up at 7 a.m. and your trip involves travelling west, covering six time zones, then on the first day at your destination you should aim to spend time outdoors between 7.30 p.m. and 10.30 p.m. before avoiding light until 2.30 a.m. or later. On the second day, the dose of bright light should come between 10.30 p.m. and 1.30 a.m. and light should then be avoided until 5 a.m. or later. Doing this slims down the number of days taken to adjust from around a week to three or four days.

The future of the twenty-four-hour society

Depending on how the market for over the-counter pharmaceuticals develops, we may see widespread use of an increasing selection of drugs which maximise our capacity for round-the-clock living. Throughout the most intense periods of the Iraq invasion, Modafinil was used by British and American troops. This allowed them to continue operations for forty-hour stretches at a time. In *The 24 Hour Society* Leon Kreitzman talks about his vision for a future where sleep is optional: 'We spend one third of our lives asleep. If we eliminated that we'd have another 25 years to do things.' If Kreitzman's reality turns out to be spot-on then future surveys of time use of the kind I talked about in the last chapter will be able to discard the category of sleep altogether. But surely a sleepless society would make us a population of robots, where time references like tomorrow, today and yesterday would be meaningless?

The UK's Future Foundation shows that whereas 35 per cent of people with an annual income of £46,000 or more use out-of-hours services, only 20 per cent of those earning less than £10,500 do. View this statistic alongside the conditions of the migrant workers in China and the Sivaskian child labourers I talked about in the last chapter and we see that the vast majority of round-the-clock work is carried out by workers at the lower end of the socio-economic hierarchy and that people who are cash-rich have the temporal upper hand in terms of which portion of the clock they choose to colonise.

The Future Society projects that by 2020 a quarter of the UK population – thirteen million people – will be working between 6 p.m. and 9 a.m., compared with seven million now. No one has yet been able to attach a precise risk factor to the long-term effects of working through the night, but if, as has been said, the effect is equivalent to that of smoking twenty cigarettes a day, then we could be talking about a cost of around ten years of life. Given that, worldwide, the number of people aged over sixty represents 10 per cent of the population and in the next fifty years is expected to rise to 20 per cent, it is likely we will see more older workers doing shift work. As I showed earlier, older people typically have less light entering the eye, and therefore may have more problems adjusting to night work. Can we continue to consume round the clock while disregarding the long-term health of the workforce? The impact of night work and other irregular-hours work needs urgent consideration alongside other factors known to affect longevity like diet and activity levels. It is these influences on our health – as well as other, more surprising ones – which come under the spotlight in considering our next relationship with time, that of individual longevity and the occupation of time by multiple generations.

Chapter 3

TWO AND A HALF BILLION HEARTBEATS: YOUR LIFE IN TIME

Imagine that today is your 122nd birthday. Your family lay on a party. For the first time ever, you get to meet your ten-year-old great-great-great-great-great-great-granddaughter. Not only that, but you are also reunited with all other seven intermediate family generations. Journalists jostle for a scoop and plague you with questions about both the secret of your longevity and what it is like to live as part of a nine-generation family. After evading death by interrogation, you survive another day and use what limited vision you have remaining to read the newspaper headlines: 122-YEAR-OLD IN NINE GENERATION PARTY. All that now stands between you and the current world record are 16.5 million heartbeats. Another four months, you will have survived longer than Jeanne Calment, who died in France on 5 August 1997 at the age of 122 years and 164 days.

Whether being a member of a family comprising nine living generations is also a record is more difficult to determine. Households with three or more generations now make up 4 per cent of the USA's 105.5 million households, which represents a growth of 60 per cent over the last decade. However, what this figure clearly does not reflect are any 'supergenerational' families living under separate roofs. While records for individual longevity are meticulously kept, there appears to be no official register keeping track of families with exceptionally large numbers of generations. Back in the 1980s, Linda Burton, a social scientist, reported an isolated case of a seven-generation family. She found the family living within a few blocks of each other in west Los Angeles.

The eldest member of the Baker family was Elizabeth, aged 92, followed by her daughter Laura, 77, granddaughter Emilia, 59, great-granddaughter Jill, 44, great-great-granddaughter Angela, 30, great-great-great-granddaughter Lisa, 15, and Lisa's son (and Elizabeth's great-great-great-great-grandson) 10-month-old Michael. The spread of ages shows that with the exception of Laura all the women had their first child aged fourteen or fifteen, within a couple of years of childbearing first being physically possible – at least according to world norms. The Baker family therefore came spectacularly close to having the maximum number of living generations possible within the breadth of Elizabeth Baker's nine decades.

In different ways, both Jeanne Calment and the Baker family are examples of humans entering previously uncharted territories of time. In Jeanne Calment's case, living the longest officially recorded life earned her the status of having the longest-known individual human relationship with time. Despite being born at a time of high infant mortality, poor obstetrical care and lack of immunisation programmes, she overshot the average life expectancy for people born in the late 1800s by some *seven* decades. Even against the backdrop of today's vastly improved life expectancies, Calment lived fifty-seven years longer than today's global average. While her achievement reminds us of the human potential to occupy a large expanse of individual time, the Baker family's story illustrates how humans also have the potential to occupy time collectively. In the 1980s five-generation families were still pretty rare in low-mortality countries – so a seven-generation family like the Bakers really broke the mould.

Although the Baker generational cascade – call it a familyscape – is remarkable enough, it is actually not a patch on what is technically possible. If we create an (admittedly unlikely) familyscape based on the world longevity record and use generational spacing based on the age of the world's youngest mother, the full extent of human potential for both the individual and collective occupation of time is revealed.

Human potential for the occupation of time: individual and collective

The world's youngest mother was a Peruvian called Lina Medina from the Andean village of Ticrapo. Incredibly, and disturbingly, Lina was aged only five years, seven months and twenty-one days old when she gave birth in 1939. Medical records showed that she had been having a regular cycle since the age of three, and that her ovaries were functioning like that of a fully mature woman. No one is aware of who her baby's father was. A child psychologist observed that Lina related to her son Gerado as if he were a baby brother rather than her son, and preferred to play with dolls rather than him. This outcome was rather unexpected given that many young girls take on the maternal role when their mother leaves or dies.

A familyscape based on the world longevity record and Lina's age – rounded up – would run from 122 through a female line of 116, 110, 104, 98, 92, 86, 80, 74, 68, 62, 56, 50, 44, 38, 32, 26, 20, 14, 8 to 2. The aggregate derived from totting up the ages from across these twenty-one generations is 1302, which represents the largest amount of cumulative time that generations living this century could technically occupy – until that is, someone outlives Jeanne Calment's 122 years or someone has a child younger than Lina Medina did. Since looking at the likelihood of a new longevity record is far less morally dubious than looking at the likelihood that someone would violate a sexually precocious child's right to a childhood, I now look at the former.

According to the register of the world's oldest people, that is those over 110 years, maintained at the Los Angeles School of Medicine, University of California, the number of people on the register has remained static at between thirty-five and forty-five over the last four years. However, we know that because the database is volunteer-run, only around 10 per cent of the world's population ever gets picked up. Most of the supercentenarians are female, replicating the survival advantage women have over men at younger ages, an issue which gets another airing later on in this chapter. Although the majority reach 113, survival past 115 is extremely rare and the nearest contenders are still some seven years off Jeanne Calment's record.

In the absence of the promise of any new world record, scientists continue to speculate about the future of longevity. Steve Austad, a gerontologist from the University of Idaho, is convinced that living to the age of 120 completely understates human potential. Austad has made a bet with Jay Olshanksy, a biodemographer from the University of Illinois, that by the year 2150 someone will live to 150. They agreed to put $150 each into an investment fund, so that if someone is officially verified as living to 150, Austad's progeny will enjoy a windfall of around $500 million. Whereas Austad sees recent improvements in longevity as part of a general ongoing trend, Olshanksy argues that unless there are interventions which alter the biological mechanisms of the ageing process, even in countries with adequate health systems we are already approaching our limit. This is because these countries have already undergone a dramatic reduction in deaths due to childhood infections, parasitic diseases and complications in childbirth. Given that the oldest living person is currently 115 years old we now have to wait until at least 2012 – and maybe substantially longer – to find out if any of the current supercentenarians will outlive Calment's record.

The scope for humans to influence the time of their life, and that of future generations

One of the things which makes humans unique as a species is that they are able to think about their future and their past lives. Aged thirteen at the junior football club, you fast-forward your mind to the day you will play for your country. At twenty-five, you save a bit of cash to allow you to go travelling the following summer. Aged fifty-eight, although you still picture yourself scoring the winning penalty in the World Cup (though possibly with a visually enhanced mental image of yourself), you also start planning how you will spend your retirement. Not all humans can do this. Towards the end of the book, I talk about why people diagnosed with autism are generally unable to relate to time in the same way.

That humans have the ability to mentally project across their entire lifespan is particularly relevant in the twenty-first century when there is a steady stream of research findings relating to longevity. Those of us living in low-mortality countries have the luxury, then, to adopt the latest anti-ageing tactics.

Another way in which humans consciously think about their individual lifespans is in terms of reproductive capacity. Again, as we shall see in this chapter, this not only differs between people but appears to vary according to the part of the world they live in.

Earlier I showed that we can conceive of human relationships with time in terms of the degree of spacing between family generations. This also happens to be particularly timely in the context of the new century. As I show later in this chapter, some surprising findings are being made about the relationship between women's *and* men's ages and the amount of time they choose to leave until becoming parents.

Now that I have sketched out the various relationships humans share with time both within and across the course of their lifespan, let's home in on the detail, starting off by looking at individual longevity.

A brief history of time: the case of human longevity

In Rome 200 BC, the average life expectancy at birth was eighteen. In eighteenth-century France, the average was thirty-five. Two centuries later, in developed countries, it stood at forty-seven. At the start of the twenty-first century, the world average life expectancy is sixty-five years. Just how much deviation is there from this figure?

In sub-Saharan Africa, where as many as one in three people is infected with the HIV/AIDS virus, the average life expectancy at birth has fallen to around the mid-thirties. And there is worse to come. Figures from the US Agency for International Development estimate that by 2010, eleven countries in sub-Saharan Africa will see life expectancies plummet further. Take Botswana: by 2010, the average life expectancy is projected to be as low as twenty-seven years. In many rural locations where around one-third of women aged fifteen to nineteen will have begun childbearing, a typical family from Botswana will in future only comprise two generations or shrink to a single generation. Those infants who do not die will spend much of their childhood caring for dying family members, while remaining at high risk from infection themselves. In this scenario the clichéd question, 'What are you going to be when you grow up?' is redundant.

It is ironic that antiviral medication, a key factor in increasing longevity in these areas, is available at a subsidised cost of $10, less than many popular anti-ageing cosmetic creams available in the UK and US. The destiny of future generations in sub-Saharan Africa lies in the hands of the international community. The secretary general of the United Nations, Kofi Annan, estimates that $7–10 billion a year is needed to fight AIDS, tuberculosis and malaria. Compare this with the size of research budgets in low-mortality countries aimed at unravelling the mechanisms of ageing and nudging up longevity. The annual budget for the USA National Institute of Ageing is nearly $50 million. In Japan, the cost of setting up the National Institute for Longevity Sciences was US$140 million, and this is topped up by an annual grant from the Ministry of Health and Welfare in the region of US$25 million.

Compare Africans' abbreviated relationships with time to those of citizens of countries at the other extreme. The Japanese government has invested heavily in longevity research programmes, so it comes as little surprise that with an average life expectancy of 81.9 years, Japan tops the league table. Not far behind Japan are various countries in western Europe, North America and Australia, with the UK standing at 78.3 years. Projected life expectancy in Japan in 2050 is 88.3 years. As well as leading the world in terms of having the highest average life expectancy, Japan also tops the league in terms of the highest healthy life expectancy. A baby born in Japan today could spend 74.5 out of its eighty-plus years with good mobility and decent health.

It now seems that life expectancy at birth will creep up enough to disprove Jay Olshansky's view that it is unlikely to exceed ninety years before 2150 – at least on a country-specific basis.

Having charted the stark differences in individual longevity, let's look within the human lifespan to determine how much variation there is worldwide in the size of the available time window to have children, and identify some of the reasons for differences.

World differences in the size of the time window to have children

Poor nutrition alters the body's composition in terms of the ratio of lean mass to body fat and delays the onset of a girl's first monthly cycle (menarche). Between the nineteenth and twentieth centuries there was an average reduction in the age for the onset of menarche by a whole year, explainable by an improvement in diet and general living conditions. In the case of British teenagers, things have evened out since the early 1950s; there has only been a negligible decrease in the average age of onset. The average age for the onset of menarche in Europe, the USA, Japan, Australia and New Zealand falls at between twelve and thirteen and a half years.

In the first chapter I mentioned that 20 per cent of the world's children had never been to school, and that 80 per cent of the rest of the world's population only spend three or four years of their lives in education. Higher illiteracy rates are linked with a later menarche. If children do physical work all day they expend more energy than they would in school. In girls this energy debt – together with any nutritional shortfall – influences ovarian function, which delays the onset of their first cycle. I also mentioned the plight of Sivaskian child workers who stay awake half the night, which is also likely to disrupt the cycle. In Cameroon, Yemen, Somalia, Nigeria, Tanzania, Haiti, Bangladesh, Papua New Guinea and Senegal the average age of the first monthly cycle falls between fourteen years five months and sixteen years two months.

What about variability in the timing of the menopause? In Nigeria and Ghana the average woman reaches menopause at the age of forty-eight compared to the USA, France, Finland, Sweden, the Czech Republic, Switzerland and Australia, where the average falls somewhere between fifty and fifty-two. Of course, in places where the average life expectancy in young adults is curtailed, a woman's window of opportunity will also be shorter.

World differences in the timing between generations

Of the fifteen million babies born to teenage mothers worldwide each year, about eight in every ten are born in sub-Saharan Africa. Once girls start their cycle, even if they attend school, the expectation is that they should leave to get married and have babies. In poor rural communities where girls reach menarche up to three years later than in developed countries, this very slightly increases the spacing between generations.

Large inequalities exist in the extent to which women are able to organise the timing of new generations. Currently around six out of ten women in sub-Saharan Africa and three out of ten in Latin America and the Caribbean are unable to access contraception. In Ghana, 50 per cent of women aged 15–19 want contraception but are unable to access it. As a result, families comprise more compact generations with large numbers of children. In Somalia, the average number of children is seven per woman, whereas in Pakistan families of twelve children are not uncommon. In countries where contraception has become more widely available, there are some signs of a drop in the fertility rate. Take Tanzania, where some eighteen years ago the number of children per family was 4.8 for urban and 7.1 for rural areas. Now it is down to 4.1 and 6.3 respectively. However, although there is demand for family planning, there are still problems accessing it and, as a result, the risk of sexually transmitted diseases is high. Research shows that even if all the available supply of condoms was distributed in developing countries then this would allow couples to have protected sex only three times a year. Whether or not this constitutes a sexual drought, I'll leave to you to decide! Part of the problem is that people are not mobile enough to get to a clinic, especially in rural locations. Organisations like Marie Stopes campaign for the right of women to determine the size of their families, and are critical in initiating change in this area.

Despite rising longevity in Japan, USA, western Europe and Australia, it remains unlikely that future weddings, thanksgivings, bar mitzvahs and other family gatherings will involve more than four generations. All the signs are that relatively recent trends in delaying motherhood and treating early teen parenting as an untimely life event lead inevitably to generations being more spaced out than in countries where other values are held.

Forty years ago the contraceptive pill first became widely available and women started to exert greater influence over the timing of the next generation. Now, with more women staying in education, the gap is widening further as they postpone having their first child until their late twenties or early thirties. In the USA the number of women of childbearing age delaying motherhood for education and a career has now reached nearly 44 per cent, representing an increase of around 10 per cent since 1990. The average age of first motherhood in Britain is now the highest in Europe, having risen to twenty-nine; in fact, a thirty-five-year-old degree-educated woman on an average income has only a 37 per cent likelihood of having a baby by the age of forty-one. In Britain this finding is causing some debate about whether sex education classes should include discussion of when to schedule in having a baby.

Postponing having children is contributing to a decrease in family size. In the UK the rate is down to 1.6 children per family – the lowest since records began in 1924. In Japan the fertility rate is also the lowest it has ever been, down to 1.28 children per family. This latter decline is due to people marrying later, as well as the rise in the proportion of Japanese women who never marry. According to the Japan Pet Food Association, dogs are replacing children as the new generation. The number of dogs – especially chihuahuas – is growing much faster than Japanese women are reproducing. In Italy and Hungary, the birthrate is down to 1.2 children, which has repercussions for an increasingly isolated elderly population. Widowed and not in contact with his only daughter, former Latin teacher seventy-nine-year-old Giorgio Angelozzi placed an advert in *Corriere della Sera*, asking to be adopted as a grandfather. The same week he was flooded with requests and took his pick. He is now looking forward to a new life with his adoptive family.

In Japan, care for the older generations was traditionally the duty of the daughter-in-law; sons and daughters who put their parents in nursing homes often faced ostracism. In 1987, when life expectancy was marginally lower than it is now, the daughter-in-law cared for an ageing relative for an average of four years. Now this has leapt to ten years and care is typically spread beyond the daughter-in-law across a wider number of relatives – between daughters, sons, spouses and other family members. In an effort to anticipate the consequences of an increasingly ageing population, companies like Sanyo and Matsushita are starting to explore the possibility of robots providing household care for older generations. It remains to be seen how long we'll have to wait to see robots take the Chihuahua generation for a walk!

The trouble with generalising trends within countries is that it runs the risk of missing out on 'subcultural' differences which influence the timing of the next generation. For instance, as well as meeting the Baker family in Los Angeles, Burton also met twenty other multigenerational black families living in Gospel Hill, a semi-rural community in the north-east United States. In these communities Burton found that many activities typically associated with mothering in most of westernised society were carried out by grand-mothers. These women placed such a high value on grandmothering that they did not consider themselves fully fledged parents until they became a grandmother. One thirty-five-year-old said; 'I suspect that my daughter [fourteen years old] will have a baby soon. If she doesn't I'll be too old to be a grandmother and to do the things I'm supposed to do, like raise my grandchild.' As the grandmothers rather than the mothers were responsible for doing the bulk of the child-rearing, they wanted their daughters to have a child early on so that they would still be physically able enough to get fully involved. In this case, the expectations and values of an individual generation and subculture determined the size of the gap between generations.

Variation in the timing of the next generation depends on access to contraception, cultural values, access to education as well as career opportunities. As I show at the end of this chapter, future developments in reproductive technology may also intervene to offer women more scope in prolonging the gap between generations.

So far I have looked at longevity and generation-related aspects of time in isolation from each other, but it would be useful to consider the relationships between the two.

Linking the age of becoming a parent with eventual longevity

When Thomas Perls from Harvard Medical School looked at the reproductive profiles of a group of women who had celebrated their hundredth birthdays compared with those who had died at seventy-three, he discovered something rather fascinating. Although all were born in 1896 and had shared similar standards of living, the centenarians were *four* times more likely to have had children after the age of forty than the women who had died at seventy-three. One possible explanation for this difference in longevity could be that the biological factors conducive to 'middle-aged' motherhood might also be associated with a slower pace of ageing and subsequent longevity. Although more research is needed in this area, Perls suspects that it is not the delay in the onset of menopause itself that leads to increased longevity – rather the hormonal conditions associated with breastfeeding, giving birth and childbearing may all positively influence health and survival in middle-aged women. This hormonal involvement also relates to the issue of why women are thought to outlive men. Women's oestrogen-rich body environment is thought to play a strong determining role due to its protective antioxidant properties, while a second factor relates to the fact that women's monthly cycles go on for some forty years – as a result they tend to be more deficient in iron than men. A reduction in iron in the body lowers the production of free radicals, which means less work for the female body in mopping up these harmful chemicals.

One extremely puzzling finding which does not fit with these biological explanations comes from Zeng Yi and James Vaupel from the Max Planck Institute for Demographic Research, who found that becoming a father relatively late in life is also linked with enhanced longevity. This can only mean that social factors must also be responsible. But no one yet can explain this finding. One possible explanation is that older couples might be more inclined to live healthier lifestyles and take fewer risks in the knowledge that they are responsible for young children.

The influence of age of reproduction on longevity presents only one way of tapping into the relationship between the generations and longevity, but we also need to consider how one generation can affect the next generation's survival.

Tracing patterns of adult survival back to the very beginning

Mothers who are underweight due to the unavailability of food as a result of floods, war, inflation, famine or other reasons are at risk of foetal malnutrition. This in turn increases the risk of stillbirth or mortality soon after birth. Even babies that survive are likely to be underweight and stunted in growth. And the effects do not stop there. The new generation suffers a range of negative consequences on their future relationships with time – a shorter reproductive lifespan, a lower age at menopause and ultimately reduced survival. In the annual wet season in Gambia, with reduced food availability, women lose between two and four kilos of body weight, which leads to malnutrition in their developing foetuses. With this history, the children's immune system is compromised, which makes them ten times more likely to die prematurely in young adulthood.

What about negative input from the father? Pesticides, tobacco smoke, air pollution and lead have all been shown to cause DNA strand and protein damage. Affected sperm can have long-lasting or permanently detrimental consequences for children, including birth defects, cancer, chronic disease or infertility.

Once the combined effects of maternal and paternal exposure to environmental hazards are known, we will be better placed to understand the links between them and childhood health and survival through adulthood. Such understanding will need to take into account the role genes play in enabling longevity.

What role do genes play in enabling longevity?

Caleb Finch from the University of Southern California and Rudolph Tanzi from Harvard Medical School claim that 20–30 per cent of variability in life expectancy is down to genes, with the remainder dependent on lifestyle. Perls points out that these findings are restricted to people living into their early seventies, and so do not necessarily apply to the exceptionally old. In any case, the difficulty with Finch and Tanzi's conclusions is that they are based on a technique which relies on making a comparison between identical twins, whose genes (or more precisely, alleles of these genes) are identical, and fraternal twins, who have only half their genes in common. The method is to pool resulting ages within identical and fraternal twin groups, the rationale being that if there are more longevity outcomes common to identical twins than fraternal twins, this suggests a greater degree of genetic influence in explaining variation. Although on the surface this sounds like quite a neat experiment, there is actually a serious difficulty with the technique, stemming from the different ways parents treat twins in their home surroundings. In the first three years of life, regardless of whether twins are identical or fraternal, parents typically give the pair the same kind of diet, similar opportunities to exercise (baby gym, walks in the park, etc.) as well as transmitting values about health and staying fit. As they grow up, however, depending on whether the twins are identical or fraternal, there is increasingly different treatment. The sheer physical similarity of identical twins means that they tend to get treated more similarly than fraternal twins, even same-sex fraternal pairs. One fraternal twin may develop a reputation as the couch potato, while the other may be increasingly distinguished as the sporty one, often out on their bike or playing netball. The twins are likely to choose friends who share their interests too. Whereas identical twins often continue to socialise together doing similar activities, fraternal twins are more likely to be in different friendship groups trying out different things – smoking or not, drinking or not, and so on. If the net result is that, overall, identical twins are treated more similarly than fraternal twins then, as Ken Richardson and I have argued in our work, existing estimates of 20–30 per cent of genetic involvement may be overstating the role for genes. This topic receives full coverage in Jay Joseph's book *The Gene Illusion*.

So far, I have looked at the role of genetics in people in their seventh decade, but this tells us very little about the role of genes in determining exceptional longevity. According to Thomas Perls, one way of approaching this question is to ask whether or not centenarians are a recent phenomenon. If they are, then we should be looking towards accounts which assign the environment a large role in explaining why centenarians live so long. Although back in the 1400s there are reports of famous people living almost until their nineties, like Michelangelo (1475–1564) who lived to age eighty-nine, reliable records of centenarians were not available until after the 1860s. In the last half of the 1800s Denmark typically had only one centenarian per million inhabitants. In the 1950s this number grew to five per million. Now it stands at 100 per million. A similar pattern of increase has been reported for other developed countries, with the number doubling every ten years. In 1900, 0.03 per cent of all Americans were centenarians, now the figure is 1.5 per cent. Over recent years, in Japan the number of centenarians has doubled every five years. Worldwide, the number of centenarians is increasing at the rate of around 8 per cent each year. According to Perls's argument, although centenarians were certainly much rarer in the past, the fact that they did exist means that the role of a significant genetic component in exceptional longevity should not be ruled out.

What is known about genes that may be responsible for enabling longevity? When Perls's team was recruiting centenarians into the New England Centenarian Study, they started to notice that the brothers and sisters of the centenarians also lived to exceptional ages. As siblings not only share 50 per cent of their genetic material but also their early home environments, individual health behaviours and living standards, all of these could potentially play a role in explaining a high prevalence of exceptional longevity in families. Looking at the survival of brothers and sisters of 102 centenarians, Perls found that, compared to the general population, brothers of centenarians were seventeen times more likely to reach 100, and sisters were at least eight times as likely to do so. Perls also found strong evidence of parents of centenarians achieving substantial longevity compared to life expectancy at the time. Although both of these findings show that exceptional longevity does appear to run in families, this could argue as much for a role for genes as it could for the diet and lifestyle of the family. What is unknown is how genes and environment combine to enable centenarians to conquer diseases or avoid them in the first place. One argument is that if the survival advantage of siblings is due mainly to environmental factors then we would expect to see an advantage which should decline with age after the siblings leave home and adopt different healthy or unhealthy lifestyles. As there is no evidence yet that this is actually the case, we have a base from which to assign genes a significant role in determining exceptional longevity.

The job of hunting for longevity-enabling regions on chromosomes falls to molecular geneticists. They not only have the challenge of trying to work out how individual genes act under their own steam but also need to piece together how genes interact. Potentially, there are so many gene interactions that there are currently insufficient centenarians to represent each permutation. Future progress in understanding the genetic contribution to exceptional longevity will depend on cooperation between researchers across countries.

Perls points to the involvement of two different classes of genes likely to play a role. First, centenarians are likely to lack a 'disease gene'. For example, they have often been found to lack an allele associated with Alzheimer's disease and cardiovascular disease. A second possibility is that centenarians possess a number of genes which work to slow ageing. Perhaps the most exciting discovery to date comes from Annibale Puca and her team from the University of Howard Hughes Medical Institute. They looked at 137 pairs of siblings who had lived to an exceptional age, and uncovered a portion on Chromosome 4 that is strongly suggestive of genes which promote exceptional longevity.

With the explosion in the number of centenarians, there are more chances than ever to study the interaction between genetic make-up and other factors that work together to enable extraordinary longevity. The United States 1990 census showed that centenarians were actually *more* likely to be poor, widowed, and to have fewer years of education. This counterintuitive finding shows that living in less than optimal conditions does not stop us reaching 100.

Although we can't choose our parents or determine the conditions we live in during childhood, when it comes to lifestyle choices we have options – but does our lifestyle really make that much difference?

Positive impact of lifestyle

If I had known I was going to live this long I would have taken better care of myself.

– 110-year-old Hermann Doernemann,
the oldest man in Germany in 2003

What lifestyle factors do centenarians attribute their exceptional longevity to? In the last two years of Jeanne Calment's life she was known to tuck into a daily slice of chocolate cake and a glass of port. At last, the evidence women everywhere need in order to indulge in guilt-free chocolate eating! Other centenarians asked to reveal their secret usually come up with at least a couple of reasons each. In 1989, Mrs Hughes, celebrating her 112th birthday, put it down to 'a good honest life' and 'adherence to the Ten Commandments'. In 2001, Amy Hulmes, Britain's oldest woman (who lived to 114), recommended four bottles of Guinness every night and cold baths. Others have put their longevity down to 'luck' and the 'good Lord'.

What about male centenarians? Retired farmer David Henderson lived in Kincardineshire, Scotland and was the UK's oldest man until he died just short of his 110th birthday in 1998. He was also the oldest person to have a pacemaker fitted. He attributed his long life to a daily bowl of porridge, never going to bed on a full stomach and a daily dose of cattle salt mixed into a shot of gin. Antonnio Todd (aged 112) from Sardinia – one of the world's centenarian hot spots – declared his recipe for longevity was to 'just love your brother and drink a good glass of red wine every day'. The beneficial effect of social ties and the antioxidant properties of red wine get a mention later in the chapter.

Until his death in 2000, aged 109, Harry Halford was the UK's oldest man. He smoked until he was sixty-five and his secret for a long life was to start each day with a cooked breakfast. His exceptional longevity appears at odds with his earlier smoking history. Laura Tafaro and co-workers at University LaSapienza discovered that out of 157 centenarians living in Rome, over 83 per cent had never smoked, 13.5 per cent were former smokers and 3.5 per cent were current smokers, suggesting that former smokers and smokers like Halford are in the minority among those living to over 100.

These accounts show that most centenarians attribute their advanced years to their dietary habits. Despite living in the era of the Human Genome Project, none of the centenarians referred to having 'good genes'.

If you are looking for evidence that positive lifestyle choices have an effect on individual longevity then you should be convinced by evidence from Gary Fraser and David Shavlik at Loma Linda University in California, gained from surveying Seventh-Day Adventists. It shows that optimal behaviours in diet, exercise, smoking, body weight and hormone replacement therapy in combination can give you the luxury of ten years' extra life.

Are these the very same lifestyle factors that explain why Japan has jumped to the top of the world's longevity table? William Cockerham and co-workers from the University of Alabama were puzzled as to why there had been such a leap in expected longevity in Japan despite there being similar economic conditions and access to a similar quality of healthcare in other countries. What these researchers noticed was a variation in life expectancy within Japan itself. In 1996 the average number of centenarians per 100,000 was 5.8 inhabitants in Japan as a whole compared to an astounding 22.1 in the prefecture of Okinawa. Not only that, but Cockerham noted that longevity was highest despite the area having the lowest per capita income of all the forty-six prefectures and the highest unemployment. The results of a twenty-five-year study offer an insight into why Okinawans achieve the world's lowest rates of cancer, stroke and coronary heart disease to lead healthy, independent and active long lives. Bradley Willcox, Craig Willcox and Makoto Suzuki report the study in a book called *The Okinawa Program: how the world's longest-lived people achieve everlasting health – and how you can too*. The Okinawan diet involves consumption of a relatively large amount of soybeans, a low salt intake of less than three grams a day, plus plenty of fish, seaweed and green vegetables. Soybeans contain isoflavones – genistein and daidzein – which are similar in structure to human oestrogens and thought to lower the risk of breast cancer in pre-menopausal women as well as to provide a protective factor against cancer of the colon and prostate.

Another outstanding feature of the area discussed by Suzuki and the Willcox brothers is 'Okinawa time'. In day-to-day life, unlike western Europeans and the rest of Japan, the Okinawans have little sense of urgency, seeing no need to rush. Although it is not yet possible to say how much a slower pace of life contributes to longevity, it is telling that Robert Levine found some signs of a link between pace of life and heart disease. But given that Japan bucked the trend in having one of the lowest rates of disease, despite being the country with the fourth-fastest pace of life, this shows that the story is not a simple one.

One characteristic of a relaxed pace of life is the chance to build meaningful social ties. Strong social connections characterising communities play a buffering role in protecting the immune system and this in turn is thought to protect against cancer. In Okinawa social ties to both the past and the present are valued – many Okinawans can trace their roots back 500 years, and spend time finding out about the origins of their family. There are various festivals that celebrate different kinds of family relationship. One sacred festival called Obon celebrates visits of the spirits as they come to live among family for the duration of the festival. Then there is Umachi – the 'festival when rice sprouts' – where people think about their living family. The Okinawans also have the philosophy of Yuimaru, which values mutual cooperation between people. This applies as much to centenarians who are respected as elders, as to other age groups.

In line with Japan's reputation for being the world leader in longevity, the government has invested in the National Institute for Longevity Sciences, creating twelve research departments equipped with forty-seven laboratories to identify the mechanisms which slow ageing. Their work includes manipulating the diets of rodents to find out if there is any relationship between food intake and average life expectancy. Mice whose food intake was reduced to 65 per cent of that of mice allowed to eat unlimited amounts of food were found to live an average of over four months longer than the controls. Although there is more than an outside chance that calorific reduction will increase lifespan in humans, to date no one has been able to test this. But what is recognised is that cutting calorific load later in life is a less effective strategy than doing so earlier. However, as previously mentioned – in relation to the longevity of Gambian women and their offspring – there are limits to this. In any case it is questionable how tolerable people would find trying to reduce their food intake, while at the same time aiming to get full daily nutritional requirements.

The *World Cancer Report* claims that 30 per cent of cancers in the West can be attributed to poor diet. Although it is now old news that eating raw fruit and vegetables is linked to reducing the chances of heart problems and the risk of cancer, the next task is to isolate how many years individual foodstuffs can potentially buy us in the fight against age-associated diseases such as cancer, heart disease, macular degeneration and cataracts. As well as thinking in terms of the number of years particular foods can potentially buy us, another aim is to link the number of portions needed with a reduced risk of a particular cancer. The antioxidant lycopene – responsible for the red colour in tomatoes, papaya, watermelon and guava – reduces the risk of prostate cancer. Consuming ten portions of tomato products (e.g. tomato sauce, tomatoes and pizza) a week has been shown to reduce the risk of prostate cancer by 35 per cent.

Then there is the magnificent antioxidant power of green tea. Naoko Sueoka and co-workers from the Saitama Cancer Research Institute have shown that if you can get through a marathon ten cups a day – about two grams of green tea extract – this can add six extra years to life for women and slightly less for men. While visiting Kyoto's Golden Temple I took part in my first tea ceremony. Having only drunk green tea from tea bags, the taste of this pure thick green product came as a bit of a shock. In the interests of international diplomacy, a speedy gulp was required before making further attempts to acclimatise to the bitter taste.

Which fresh foods contain the most antioxidants? Rats fed diets of blueberry, strawberry or spinach extracts showed signs that all of these foodstuffs improve performance on various learning tasks but that overall blueberries have the most powerful effect on coordination, balance and learning. Blueberries contain extremely high levels of antioxidants – one of which, anthocyanin, is responsible for the red, purple and blue colour of fruit – which aid cell protection by neutralising damage to cell membranes and DNA.

Apparently, munching a cup of blueberries a day will potentially slow the ageing of our brains; it is the anthocyanin concentrated in the skin that works the magic.

Whatever effect antioxidants have in isolation, Britain's Institute of Food Research has shown that their effect can be a colossal *thirteen* times more potent when combined with other anti-cancer foods. David Heber's work at the UCLA Center for Human Nutrition in Los Angeles has come up with a shopper-friendly way to monitor your intake. The trick is to fill your trolley with a rainbow of foodstuffs: purple/red (berries, grapes, red wine); red (tomatoes, red peppers); orange (carrots, mangoes); yellow/green (spinach, avocados); green (broccoli, cabbage) and white/green (garlic, onions). Fortunately, the success of this approach does not depend on combining these foods together in the same recipe.

Future scope for the re-organisation of lifespan

Advances in reproduction and fertility research may introduce new options for the timing of your own family's future. For instance, ultrasound can be used to assess ovarian volume, which in turn serves as an indicator of the number of eggs a woman has left, and provides an idea of when her clock will stop ticking. If screening for ovarian ageing were made widely available this would allow women more scope to plan their reproductive future.

Earlier, I talked about how the freezer saves time by cutting down on food preparation time, saving on shopping trips and so on. However, the role of the freezer goes beyond the kitchen. Here, we're not talking about the relatively warm climate of the home freezer. Although the use of the freezer for self-preservation after death with a view to subsequent resuscitation is an option only for a minority of us, its influence over other relationships with time is already a fact. This year saw the world's first pregnancy after ovarian transplant. A cancer patient at Cliniques Universitaires St-Luc in Brussels had the outer layer of her ovaries 'cryo-preserved'. After the cancer treatment, the ovarian tissue was replanted in her body and she later conceived and gave birth. This scientific breakthrough not only offers hope for women suffering from cancer or an early menopause, it also has the technical potential to enable menopausal women to reorganise the timing of their lives. Many women complete their education in their twenties and build their career in their thirties, precisely at the time when their fertility is starting to tail off. With today's longevity, given the appropriate ethical consideration, post-menopausal women may have the option to reverse any non-fertile status, giving them more choice about when to concentrate on career and when to focus on having a family. In any case, IVF treatments have already enabled both a sixty-five-year-old woman in India and a sixty-three-year-old woman from Italy to give birth. As I said earlier, humans are unique in their capacity to be able to reflect on the timing of new generations, but with widespread advances in reproductive technology, even more possibilities may emerge.

What about the scope for men wanting to organise the timing of their reproductive life? Some men treated for testicular cancer will remain infertile, but if sperm is taken before treatment and frozen, the Human Fertilisation and Embryology Authority allows it to be used for ten years, which is renewable in cases where the man is under fifty-five and remains infertile. If longevity continues to rise, surely this provides a case for storing sperm for longer time periods?

Advances in fertility technology and organ transplant allow us to think beyond time in terms of the chronological age of our entire body. With our reproductive capacity capable of being put on hold, and with liver, heart, kidney and bone marrow transplants now routine, along with innovations in artificial means for restoring vision (through the use of cortical implants, optic nerve stimulation and retinal implants), in future people may look at the age of individual body parts rather than the chronological age of the whole body.

World inequalities in relationships with time of life

Exactly by how much we under- or overshoot the average world life expectancy of 2.5 billion heartbeats depends not only on where we live in the world, but the combination of our nutritional intake as a foetus, our genes, the age of our parents and of course our lifestyle. Although we humans are unique in our capacity to consciously reflect on the trajectory of our life, its internal organisation and its relationship with other generations, it is primarily those of us who live in low-mortality countries who have the best chance to act on this information. By adopting optimal health habits, we have the potential to prolong our lives by as much as ten years. Being female may bump up your expectancy, so may being an older mother. The rise in the number of women in low-mortality countries postponing motherhood until their early thirties means that we should be in a position to be able either to refute or support the latter idea in around seventy years' time.

However, whatever is *technically* possible for the onset of widespread longevity needs consideration alongside world resources per capita. Take water. Droughts in Ethiopia, legal disputes in relatively water-rich North America, floods in Mozambique, not to mention rising sea levels due to global warming, collectively threaten the available land and access to safe water. By 2025, two thirds of the world's peoples are likely to be living in areas of acute 'water stress'. Already, we are beginning to detect the impact of wet seasons, water and food shortages on reproductive conditions, birthweight and future survival rate. Together these global issues have the potential to affect our individual and collective relationships with time.

While average life expectancy continues to nudge upwards in Japan, Europe, North America and Australia, what remains unknown is the effect of widespread obesity over the next century. In the United States, one third of people are obese and 75 per cent of people are overweight, and so we might see a reversal in the recent trends. At present a gulf of nearly sixty years separates the life expectancy in these countries with those in sub-Saharan Africa, where in the past, the expectation was that the grandparent generation would be cared for by the youth. Now, entire generations are being wiped out so that there are not even parents to do the caring. Given the relatively short time that humans have inhabited the earth, the human capacity to live alongside more than three surviving generations at a time is actually only a recent phenomenon. Yet, compare this to low-mortality countries, where only a few decades ago it was unusual for an adult in middle age to have a living parent. Today's adults not only have care responsibilities for children but also multiple older generations. As I mentioned earlier, one in ten women in Britain is currently caring for someone under eighteen, and also has care responsibilities for an elderly parent. Living in the era of 'sandwich generations' multiple care roles means ever-increasing struggles to coordinate with family, friends and colleagues in time.

In high-mortality countries women enter motherhood around a decade earlier than in low-mortality countries and are more likely to have larger families. In many of the former, planning when to have the next generation is not possible as there is no access to contraception, yet contraception would enable women to reduce the likelihood of having children in poor living conditions. Instead, as a result of not being able to plan in this way, the risk to the next generation is substantial – not only in terms of infant mortality but also survival into adulthood.

Widespread longevity also has repercussions for the well-being of the exceptionally old. Although we know that securing an extra ten years in the bank is possible for those living in low-mortality countries, we know less about how living beyond 100 impacts on our mental as well as our physical health. Laura Tafaro and her co-workers at LaSapienza University examined forty-two centenarians to find that the majority showed either no or very little aggression. The most common disturbances were anxiety, depression, fear of being left alone, being suspicious and doing the same activity over and over. Alongside efforts which look at how we can continue to nudge up longevity, there needs to be more research into how best to support people who care for centenarians.

When Thomas Perls looked at mental clarity, he estimated that 20 per cent of centenarians do not show signs of mental impairment. In nine out of ten of the remaining 80 per cent who do, this does not occur until an average age of ninety-two. Those who reach 100 are relatively resistant to dementia. But even if dementia-free, surely survival to this age gives the healthy brain an organisational nightmare in terms of being able to recall experience of time over the previous ten decades? It is this particular relationship with time – our *experience* of it – which forms the topic of the next chapter.

Chapter 4

EXPERIENCING TIME: THE CASE OF THE TWENTY-FIRST-CENTURY MIND

The year is 2984. Surgically implanted into the fleshiest part of your upper arm is a bar code. Swipe this across any time-travel sensor and you have immediate authorisation to download from the World Time Bank. As your body is already installed with the necessary neural add-ons to enable transactions to occur at the multi-sensory level, you get experiences of a particularly high calibre. Today you customise your time trip by opting for a taster tour of city life in Toronto 2006. After receiving clearance from neural immigration your body undergoes a series of preparatory transformations at the neural-experiential level. Then, zap! You are transported to the 2006 neural port, where after enduring an initial introductory spiel – on Toronto's history, climate and population size – you get to experience life through the eyes of a bystander. People are suitably dressed for the early stages of global warming and spend more time being up during the day than they do at night. It feels odd that they only appear to be able to do three things at once – typically walk, swig from their take-out coffees and chatter on their mobiles. This contrasts with 2984, where in addition to consuming and talking with others your network schedule monitor (NSM) operates round-the-clock updates from your multiple life schedules as developments happen – be it robots, pets, spouse, children, multigenerational family members, friends and yourself. How on earth did people living in twenty-first-century industrialised society get by before the advent of NSMs?

This chapter has as its mission portraying the nature of people's experiences with time at the beginning of the twenty-first century. In doing this, I will recall some of the themes from earlier chapters – exceptional longevity, life in clock time cultures, and the involvement of various brain regions in timing.

The continued rise in longevity enables new experiences with time

At the start of the twenty-first century, low-mortality countries continued to witness increased longevity, which gave people – old and young – the scope to experience time in new ways. Let's now uncover why this was the case.

In Chapter 3, I talked about the uniquely human ability to mentally zip back and forth along life's timeline. As extended longevity allows a longer time path than was previously possible, this offers even more scope to experience the planning of future time. As it became no longer unrealistic for people living in the twenty-first century to survive until their eighties, they started to have the luxury of projecting their thoughts about the future over a longer expanse of time than previous generations. This was also reflected in the ways they started to refer to different chunks of life. Whereas the 'midlife crisis' had only been heard of at the end of the previous century, it became routine for people to find themselves in card shops forcing a weak laugh at the humour in cards celebrating this event. The mid-twentieth-century categories of adulthood devised by psychoanalysts, psychologists and others had become dated by the following century because they did not take into account extended longevity. Table 3 shows Erik Erikson's theory of identity development involving eight stages with approximate ages. Given that at the start of the twenty-first century centenarians became the fastest-growing portion of the population, increasing numbers of people were into the 65+ category, 'maturity', a term which did not do justice to the four decades it now covered (not to mention the potential scope for further transitions in people's identity during this extended time period).

Another impact of extended longevity relates to the speed of the approach of certain life events. Certain milestone birthdays like the sixtieth, seventieth or eightieth, or annual celebrations like Diwali or Christmas, come around subjectively more quickly as we age. But we know rationally that the same

Table 3: Erikson's stages of identity development

Stage	Ages	Basic Conflict	Important Event	Summary
1. Oral-Sensory	Birth to 12 to 18 months	Trust vs Mistrust	Feeding	Feeling of physical comfort, being able to form a first loving, trusting caregiver relationship, accompanied by minimal fear.
2. Muscular-Anal	18 months to 3 years	Autonomy vs Shame/Doubt	Toilet training	Child gains a sense that their behaviour is their own, they become attuned to their free will. Experience control but experience shame and doubt if restrained or punished too harshly.
3. Locomotor	3 to 6 years	Initiative vs Guilt	Independence	Child engages in purposeful behaviour, able to imagine as well as feel remorse for actions.
4. Latency	6 to 12 years	Industry vs Inferiority	School	Child shifts to achievement oriented approach, learning to do things well or correctly in comparison with others.
5. Adolescence	12 to 18 years	Identity vs Role Confusion	Peer relationships	Teenager develops sense of self in relation to others, develops a self identity in sex roles, religion and culture.
6. Young adulthood	18 to 40 years	Intimacy vs Isolation	Love relationships	Adult develops ability to give and receive love; long term commitment to relationships.
7. Middle Adulthood	40 to 65 years	Generativity vs Stagnation	Parenting	Adult develops interest in shaping the development of the next generation.
8. Maturity	65 to death	Ego Integrity vs Despair	Reflection on and acceptance of one's life	Adult develops a sense of acceptance of life, and value of relationships developed over lifetime.

amount of clock time has passed, so why does this happen? One tempting explanation is that time appears to speed up because, as we live longer, a particular period of time takes a smaller relative proportion of our total lived time to date. Both of these examples show how extended longevity enables new experiences in terms of thoughts about future time.

The extended longevity at the start of the twenty-first century also influenced people's experiences of past time. Although centenarian brains had the potential to recall different times in life – their childhood, teenage years, early adulthood, midlife, post-midlife as well as recent times – to what extent were they actually able to do this? Pia Fromholt and her co-workers at the University of Denmark used aids to trigger the memories of twenty-two Danish centenarians of events over their entire lives. They also carried out the same tasks with people with Alzheimer's, as well as a group of unaffected people aged eighty. Figure 3 shows the distribution of memories occurring across time.

The pattern produced by the centenarians was similar to the profile of those aged eighty, with a period of childhood amnesia, a bump upward around the age of twenty, followed by a progressive dip thereafter until the age of eighty-five, with a final spurt representing most recent memories. That there was a bump around early adulthood fits with what Costa and Kastenbaum found when they interviewed over 250 centenarians and asked them to report three types of memory. These were their 'earliest memorable event', the 'most salient historical event' and the 'most exciting event' in their

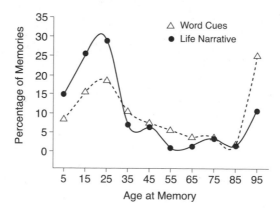

Figure 3 **The percentage of life-narrative autobiographical memories that occurred in each decade of life of centenarians.**

lifetimes. In the exciting event category, nearly half of all memories were located between the ages of thirteen and thirty-nine rather than during the sixty-year expanse between forty and 100. Given that this is the period when most people get married and start a family, this is perhaps hardly surprising. In the younger age group, half of all their recalled memories came from the previous year, whereas less than a fifth of the centenarians' memories were located in this period. That the centenarians still showed some memory of recent events – albeit less than those in the younger group – showed that they were still in touch with their latest experiences of time, yet clearly their default relationship was with earlier time periods. One possible explanation of why this might be so comes from Jeanne Calment five years before her death, 'At 117 years some days are all the same. You need an extraordinary memory.' In nursing homes and care homes, where each day is highly structured, the content of ordinary days is more likely to be forgotten and earlier memories dominate.

Another experience of time afforded to seniors is the opportunity to experience time through the vehicle of having different conceptions of the self, which appear either close or distant from previous conceptions of the self. As we age, we have a greater expanse of time behind us featuring our old selves with which we can make comparisons. The start of the year, according to the Gregorian calendar, is 1 January. This is the traditional time for shaking off last year's self and launching the 'new you' – the 'fit you', the 'organised you', the 'financially solvent you' and so on. It is not only 1 January when we distance ourselves from our previous selves. With increased life expectancy, people have more time to experience more life changes, which drive these new or reworked self-conceptions. House moves, the birth of grandchildren or great-grandchildren and retirement plans all allow transitions. People respond positively or negatively to their former self in terms of how distant that self *feels* in time. They also evaluate former selves more critically the further away they feel subjectively from them. Our experiences of ourselves in time depend on how we subjectively experience the time elapsed between life events and our identities linked with those times.

Research on Canadian and Japanese populations shows that people see themselves as akin to fine wine – as something that improves with age. As the years go by, they think of themselves as becoming progressively better versions. Michael Ross and Anne Wilson from the University of Waterloo show that at fifty people regard themselves as superior to their peers, and not only that but even better than they apparently thought they were at age twenty. In the same way, university students perceive themselves as superior to their peers than when they were sixteen. Ross and Wilson point out that while of course life experience does mean that people get 'better', the finding that people consistently perceive themselves as better than their peers was rather unexpected.

Keeping track of time in twenty-first-century clock culture

Regardless of our age, our experience of time at least partially depends on how society treats time and its passage. For those living in early twenty-first-century clock culture, it was instrumental to an individual's success to be constantly checking whether or not their back-to-back schedule was on target for its anticipated time.

Clock culture and the experience of short time intervals

In twenty-first-century clock culture wearing a wristwatch did not automatically guarantee that people no longer had to make subjective estimates of duration. Despite the importance placed on punctuality, it was regarded as inappropriate to interrupt a conversation or to appear uninterested by overtly checking your wristwatch. Some people wore the face of their watch on the inside of their wrist so that in formal situations they could find out the time inconspicuously. Another common experience was phoning someone only to have their colleague say, 'If you call back in ten minutes, you should just be able to catch her before she leaves the office for the day.' But you would always forget to look at your watch to note when the ten minutes began, so would have to rely on your judgement as to when to call again.

Another example of the passing of a short time interval is a uniquely feminine one. Summer leg exposure often called for a certain amount of preparation. Back then, one way women defuzzed was to smother their legs in hair removal cream. The sheer messiness of the operation meant leaving any watch well away from the scene. Accurately estimating the cream's statutory application period of three minutes without referring to a watch falls into the hands of one highly specialised brain mechanism. Underestimate the time and the purpose of the operation is defeated; overestimate and you end up with an angry rash.

How do humans keep track of time intervals? The answer to this partially depends on how long the interval is. Those in the range of hundreds of milliseconds to several seconds can be perceived as a single unit, whereas those longer than three seconds need to be estimated as a number of subunits such as seconds. One explanation is that we consciously count at regular intervals. But this is not something that is always practical. Besides, we seem to be able to get by without using this strategy. What's more we know from watching the responses of birds and fish that they are capable of holding off making a trained response for food from specific time intervals. Given that as yet there have been no sightings of birds making tallies in tree trunks or reports of fish doing the same in river beds, we need to look at what enables the human brain to accomplish this feat. Some scientists claim that some kind of interval time clock exists, which works to process information of a temporal nature. Different types of model have been proposed, but the most popular one posits that a switch between a pacemaker and accumulator flips to allow an internal pacemaker to produce a series of pulses. As these pulses mount up, they are counted by the accumulator and stored in memory. An evaluation is made by comparing time elapsed with pre-existing represen-tations of the time period. When the period is over, the switch flips back, severing the connection.

Any internal pacemaker appears affected by whether the information about duration is presented through the medium of sight or sound. If people see a light and hear a tone lasting the same length of time, they tend to perceive the tone as lasting longer than the light. The reason for this is that the visual clock speed is believed to be around 10 per cent slower than the auditory clock. This effect is seen in children as young as five. However, young children's memories contain less precise representations of standard visual durations, as their performance was more erratic with vision than with sound. As I indicate in the next chapter, that young children are better at sounds may reflect their history of processing sounds as early as the seventh month of gestation.

In Chapter 2, I talked about daily peaks and troughs in alertness, mental performance and wakefulness according to the point reached in the cycle of the circadian rhythm. It is now known that the ability to perceive short time intervals also varies according to this rhythm. When Kenichi Kuriyama and co-workers from the Tokyo Medical and Dental University looked at people's time perception at four points in the day – 9 a.m., 1 p.m., 5 p.m. and 9 p.m. – they found that people's ability to judge the passage of a ten-second interval grew more accurate as the day went on. The worst performances occurred in the morning, when participants tended to overshoot the length of the time interval by over two seconds, but by the end of the day the overestimate had reduced to just over a second. Performance on this task was closely related to body temperature. Given these findings, who now needs to rely on lame excuses for bad morning timekeeping – in future, simply blame your circadian rhythm!

The areas of the brain thought to be involved in monitoring this kind of timing have now been narrowed down. For instance, the counting of the pulses is thought to take place in the inferior parietal cortex. The work involved in making the comparison between the estimated duration and the reference value is thought to be completed in the prefrontal cortex on both sides of the brain. Figures 4 and 5 show the location of the parietal cortex and prefrontal cortex respectively. A study by Ignor Nenadic and co-workers from the University of Jena in Germany sought to find out more about how time is represented in the brain. In the first task, on each trial participants heard a standard tone of one second duration followed by a comparison tone which was either longer or shorter than this. They then had to decide whether the second tone was longer or shorter than the first. If the participant responded correctly the task was made more difficult by decreasing the difference in duration between the tones on the next trial. If they were unable to discriminate then things were made easier by increasing the difference in length between the tones. The idea was to pin down the threshold at which accurate performance was achievable. In the second task, the aim was to provide participants with an activity that shared the mental demands of the first task, but looked at pitch instead of timing. In this way, it was possible to carry out a 'spot the difference' between the brain pathways involved in both tasks and the ones specifically involved in timing. The researchers found evidence of a centralised processing of temporal information in the area of the brain called the basal ganglia as well as a potential involvement for the putamen. Identifying which brain areas are responsible for timing is

***Right* Figure 4** Location of parietal
cortex in the brain.

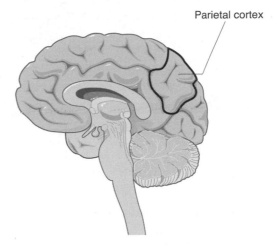

Parietal cortex

***Below* Figure 5** The prefrontal cortex.

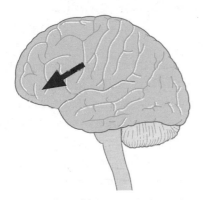

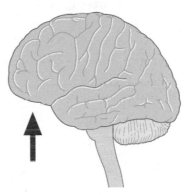

Dorsal lateral prefrontal cortex outside view

Inferior orbital prefrontal cortex outside view

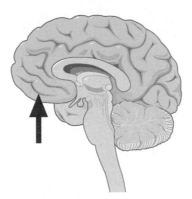

Inferior orbital prefrontal area inside view

important because, as I show in Chapter 7, knowledge is power in terms of understanding the causes and knowing how to treat disorders where disrupted timing is an issue.

The quality of the consumer wait

In twenty-first-century consumerist society, where time was money and lengthy waits for service were not acceptable, those working in customer-focused environments devised strategies to try to make customers perceive that they were not caught up in a delay. In Chapter 1, I talked about M-time cultures where the default mode is to handle tasks in an ordered sequence one at a time, with fixed appointment slots for each event's occurrence. I also commented that people from traditionally M-time cultures were more likely to perceive a wait than those from P-time cultures. In cultures with a strong tendency toward M-time, where individual achievement is of high importance, it is crucial to ensure that waiting time is kept to a minimum to increase the chance that consumers will return. Business managers know that even if there is overall customer satisfaction with the quality of the product and positive interaction with employees, lengthy queues or extended delays can ruin an otherwise perfectly good retail experience. Assuming that a customer services manager has already done everything possible to keep waiting time to a minimum, like adding more toilet booths or improving building design so that people in queues feel like they are moving, the next priority is to ensure that the *quality* of any wait is improved. In this way, customers perceive any wait as taking less time than it actually does. Although a popular trick of the trade is to play music to reduce the negativity of customer waiting, 'Greensleeves' can grate, and it grates particularly when the cost of waiting for the service is perceived as high, like missing a connecting train, or when a limited resource is perceived as running out – say tickets for the FA Cup Final. Perception of waiting depends on whether a customer is paying attention to the passage of time. If they are, their 'cognitive timer' will be fully engaged and a wait will seem longer precisely because the customer is focusing on the time that is passing.

Selective experiences relating to time

Now let's move from the general to the personal, and look at how people in the twenty-first century experienced time. You will see that some of the phenomena described are as familiar to us now in 2984 as they would have been back then. I will use examples relevant – or at least memorable – to those living in the early part of the twenty-first century. What explains déjà vu? How do people experience time while serving a life sentence? What about someone who has just been told that they have a terminal illness? How do athletes competing in timed events experience time. Lastly, we look at the puzzle of why it is easy to picture ourselves yesterday, last week, one year ago, ten years ago, but what prevents us from mentally travelling back in time to life as a one-year-old.

The case of déjà vu

Around once a year or so, 60 per cent of people experience déjà vu – a curious impression of familiarity with a particular new experience but no obvious explanation for this. As there is no shortage of theories regarding this phenomenon, ranging from the scientific to the paranormal, let's take a look at the former.

Alan Brown from South Methodist University suspects that neurological dysfunction – seizure or slow neural transmission – is not the likely explanation for déjà vu. Instead, he suspects memory or attention play a role. In the case of memory, the idea is that some aspects of the present situation are likely to be familiar but that the *source* of familiarity has been forgotten. This could arise because although we tend to process a lot of information day to day, conscious attention is not fully devoted to this process, and so it could be that a second round of processing could lead to a visual 'double take'. Put another way, a person may be taking in a scene while being only partially engaged with it. This may be followed by a second phase of attention, where the person becomes more fully engaged. However, as the second full engagement may match the previous moments, the person does not actually consciously recognise the earlier experience as being immediate but rather attributes it to a much more remote past.

Serving life imprisonment

People who society punishes with life imprisonment not only face the loss of life as they knew it but also now perceive time as controlling their lives. Diana Medlicott of Buckinghamshire Chilterns University, who interviewed lifers to find out how they experience time, summed up the nature of the punitive experience as:

> The external duration of each life has been brought to an end: henceforward the prisoners must live to prison time, unable to choose freely how to spend any time inside, and unable to participate in the chronology of events that made up their life on the outside and helped to construct and maintain their identity. Family birthdays, football matches, religious feast days, leaving parties for work colleagues – all the chronology of birth, life and death flows on the outside of the prison, and the prisoner remains bitterly aware of it while forcibly restrained from participation in it.

Despite directives stating that prisoners should be permitted a specified amount of 'purposeful' activity a day, the reality is that it is far less. 'Time passes extremely slowly. I've been in twelve days, and it seems like four weeks. Time just doesn't go.' Lifers try to overcome the pains of the present, all the while knowing that the future stores up more of the same. Medlicott distinguished between those lifers who were suicidal and those who were trying to cope using strategies – like Ken, who said, 'I telephone the wife every day and twice at weekends. We get up at 5 a.m. and write to each other. Then we have a cup of tea together at about 8 o'clock. She's changed her meal times so that we eat together at lunch and tea.' By coordinating his personal time with someone significant to him 300 miles away from prison he has gained some emotional comfort.

Time running out: being terminally ill

Some people diagnosed with a terminal illness adopt a new time perspective by consciously choosing to live in the 'here and now'. The concept of time as they knew it has been abolished – the past, present and future are transformed from three dimensions into a single one – the repetitive present. The story of

Muriel Kindler, aged seventy-one, who had surgery for a brain tumour, appeared in the *Observer*. She described her situation:

> When you get a life-threatening illness, it brings everything to a screeching halt. Everything slows down, you notice details more and have a different relationship with the world around you. It strips away the things that are not important. You learn that every day is a gift and your loved ones are very precious. It sounds like a cliché but it helps to look on the bright side.

Another experience common to people facing terminal illness is that time feels stolen. Instead of the usual routines, flexibility and schedules, hours are spent sitting in waiting rooms and undergoing medical procedures as well as coping with the reactions of family and friends. Most personal habits of timing disappear, including regular sleep patterns, so the body is out of synchrony with the light–dark cycle, which adds further fatigue to an already overburdened system.

Part of the change in relationship with time is due to it reflecting a point in our life rather than our chronological age. For instance, the perception of how much time we have left in life influences the kinds of social values we choose to take on. Laura Carstensen and co-workers at Stanford University compared life goals across men in their late thirties who fell into one of three groups: HIV-negative, HIV-positive but without symptoms, and HIV experiencing symptoms of AIDS. As these groups had similar chronological ages but different levels of life expectancy, they could be compared with people who were older in chronological age, also towards the end of their lives. The researchers found that, regardless of chronological age, when life is no longer seen as open-ended, people shift away from seeing others as potentially offering them knowledge and instead prioritise emotion as the focus of their relationships.

Time running out: athletes beating the clock

How do athletes experience time when their particular event involves the passage of short time intervals? Perhaps one landmark in the experience of time is the four-minute mile. Just before Roger Bannister ran the first sub-four-minute mile at Iffley Road in Oxford on 6 May 1954, staring him in the face was the time he needed to beat. Yet his subjective experience of this time interval on the race day gets little airing. In his book *The First Four Minutes* Bannister describes how the 3 minutes 59.4 seconds actually felt. In the first lap Bannister felt like he was going so slowly that he shouted 'Faster' to pacesetter Chris Brasher; somehow the excitement he felt during the race knocked out his sense of the pace. The first lap took 57.5 seconds, right on target. Then, at one and a half laps, Bannister reported worrying about his speed. After someone called 'Relax' from the crowd, Bannister reported that he felt himself 'let go' and reflected that he barely noticed the half-mile mark at 1 minute 58 seconds. Three hundred yards from the end Bannister described feeling like his mind was running ahead of his body and that the 'moment of a lifetime had come' – for him, the world was standing still or not existing. Perhaps unsurprisingly, Bannister described the last few seconds of the race as never ending.

Infant amnesia: a universal phenomenon?

Lastly, there is one experience of time which is universal but difficult to write about. This is because it is off-limits to us. Why are our experiences of time in infancy so inaccessible? Why is it that as adults most of us are unable to retrieve our earliest memories? One source of constraint may simply be the immature infant brain, which limits our capacity to recall memories over several years. Between the ages of eight and twenty-four months there is enormous growth in the nerve pathways in the area of the brain known as the frontal cortex. It is possible that experiences before this state of neurological readiness are not retrievable later in life.

A different kind of explanation came from Sigmund Freud, one of the first to talk about lack of access to our childhood memories. Freud regarded childhood memories – particularly sexual ones – as being too fear-inducing and repulsive to be preserved. Instead, he believed people repressed them into some more emotionally neutral form. Although this may be the case for some people, the work of Leonore Terr shows otherwise for the general population. Terr worked with children under five exposed to traumas in their first few years such as sexual abuse, witnessing death or serious physical injury. She discovered that the type of memory depended on the age at which the trauma took place. Although the children in her study may not have had any conscious memory of the trauma, all of them became highly distressed in situations that took on the surface appearance of the original episode. For instance, having another child wrap them up in a blanket at playschool would provoke a strong reaction if the child had previously been sped to hospital wrapped in a blanket as a medical emergency. Although this distress indicated that they all had some memory, albeit buried, only those children aged over twenty-eight months at the time of the trauma were able to give descriptions of what had happened. The behavioural recall of these early childhood events was not transferable into words for the younger children, even though they could talk at the time of the events. Interestingly, rats and monkeys who experience early pain or social deprivation show neurological changes in the pathway linking the hypothalamus and the body's key hormone-producing glands (the area of the brain called the hypothalmic–pituitary–adrenocortical axis) and this might also suggest similar changes could occur in the developing childhood brain.

Although adult memories of infant experiences are not accessible, this does not appear to be the case with early-life traumatic experiences which express themselves in terms of memories which the body reacts to behaviourally, rather than conscious ones we can express through speech. There is convincing evidence of substantial neurological changes which accompany the encoding of emotionally significant events. Childhood memories with a substantial emotional component – like moving house or trips to the hospital – can certainly leave their neurological mark in the brain. Martina Piefke and co-workers at the University of Bielefield asked adults to recall twenty memories from their lives before the age of ten containing two different extremes of emotional content. At the end of this phase there were ten negative childhood memories (for example, firecracker exploding near hand), ten positive childhood memories (for example, unwrapping a present to find a big Lego set), as well as ten each of positive and negative memories from their recent adult life in the five years before the interview. The idea was that by using brain imaging the subjects might be able to 'light up' those areas of the brain associated with memories from childhood and adulthood according to type of emotional experience. Piefke found that different areas of the brain's hippocampal region were involved according to the age of the memory and its emotional content. That different areas were involved suggests that the hippocampus stores memories differently according to when an event occurred – at least when those memories are emotionally laden. Memories with different degrees of emotional content and associated events at different ages were encoded in different areas of the hippocampus.

What about memories common to the majority of people's childhood experience, like the arrival of a newborn sibling? It is common for our early memories to involve siblings. My own first memory is of putting my head in my mum's lap and asking 'When is there going to be another baby?' Then the doorbell rang and an unfamiliar man appeared in the kitchen with a delivery of goldfish (apparently, in the 1970s, goldfish delivery was something that the UK postal service took in its stride). Although it is hard to pin down the exact age when I asked this question, I've narrowed it down to between the ages of two years ten months and three years. Rubin and co-workers from Duke University have looked at the conditions under which children best remember and found that those who experienced the birth of a brother or sister when aged between two years and two years three months remembered far less than those who experienced a sibling birth when aged between two years four months and three years three months. This finding helps us to locate the conditions and ages supporting successful recall of early memories. When Rubin looked at reports of memories from before age eleven in over 11,000 American adults, he found that the most common themes included birth of a sibling, the death of a family member, moving to another house and going to hospital. The memories had a peak concentration around the age of seven, with only 1 per cent of memories reported before age three. This finding is in line with autobiographies, where writers typically do not write in any detail about their lives before the age of five.

When I was five and on holiday with my parents at Weston-Super-Mare, I suddenly suffered agonising pain. I collapsed and my father had to carry me indoors. It was rheumatic fever.

Jimmy Young, *Forever Young*

I was five when the war started, and Monday 4 September 1939 should have been my first day at school, but that was not to be.

Alan Bennett, *Writing Home*

Differences between just how early people can remember may not only depend on the occurrence of salient family events but also on varying child-rearing practices attributable either to individual parental style or to cultural factors. For example, people living in rural Indian farming communities (compared to urban dwellers) did not remember any specific episodes from their past. In contrast, others have found differences in recall among cultural groups explainable by child-rearing practices. Comparing memories between different groups of New Zealanders – Maoris, Asian immigrants and people of European descent – Maoris reported their earliest memories at around two years eight months, some ten months earlier than the people of European descent and more than two years earlier than the Asian immigrants. Other studies show that Europeans and Caucasian-Americans can cast their minds back to the age of three years five months, some six months ahead of native Koreans, Chinese and overseas Asians. Qi Wang from Cornell University characterised American autobiographical memories of childhood as being generally detailed, specific, self-focused and emotionally rich compared to those of Chinese, which are typically more skeletal, generic, relationship-focused and emotionally unexpressive. Differences in parenting style may explain this. While Chinese mothers described staying close to their baby as being the trait most characteristic of their child-rearing style, American mothers viewed 'laughing easily' and 'being active' as the most typical. Wang sees these different kinds of approaches to bringing up children as contributing to the kinds of memories we have as youngsters.

Returning to 2984

Having glimpsed some of the ways in which twenty-first-century minds experience time, you now return to 2984, where time-travel sensors allows you to revisit your mental experiences whenever you like. In the age of time-travel sensors, amnesia is not an issue – unless, that is, you want to recall which particular year something took place before you could use sensors. You can't help comparing how effortlessly centenarians of the thirtieth century access their past experiences, regardless of how structured, boring or eventful their daily life is, compared to those in the twenty-first century. In the commercial environment, you notice how twenty-first-century service providers and companies are customer-focused, striving to improve the quality of any wait. Yet for thirtieth-century minds there is no need to do this as all aspects of consumerism are now portable and products are delivered within a matter of milliseconds. Scientists of the twenty-first century are struggling to explain déjà vu, but as they lack the technology, they are still some way off. Experiences like imprisonment or solitary confinement continue to be a feature of human experience of time, whatever century it is. The available technology and the brain's attunement to this alongside cultural expectations help to set parameters around the nature of our daily experiences of the past, present and future.

Now that I have toured themes around 'time of life', the next stop is to go to the other extreme by zooming right down to look at our relationships with time in one tiny fraction of our lives. Although this part of life is off-limits for most of our memories, the use of some rather ingenious techniques can tell us an unbelievable amount about how babies relate to time.

Chapter 5

TIME MANAGEMENT BY BABIES

Now really, what is your problem? After all, you know *perfectly* well what the Baby's Charter says. If I demand some milk, a nap, a hug, a nappy change, or even another round of those ridiculous farmyard sounds, I expect to have it on an anytime, anyplace, anywhere basis, and preferably not any time later than now. So why the tardiness – not to mention grumpy demeanour – in responding to my request for feeding at 3.10 a.m.? Even when I increase the status of my night terror alert from amber to red by screaming the house down, it seems to make no difference to you. With all due respect, the performance targets for response speed laid out in the Baby's Charter are not being met.

Hmmfph, I suppose you want a chance to tell your side of the story. Well OK, yes, I admit that babies have a notorious reputation for making round-the-clock demands. Around about a third of us continue night-time howling, cot-guard rattling, feed-me-*now*-or-else crying fits and other sleep disturbances beyond the first year. In the UK, the estimated cost to the National Health Service of persistent, unexplained crying in babies is £65 million per year. Even for those of us who manage regular sleep–wake cycles by the age of six months, out of all the relationships we share with time, the synchronisation of rest cycle with darkness is the slowest of all to develop. But what babies lack in the sleep at night department, we certainly make up for by our early proficiency in a whole host of other relationships with time. Before putting the spotlight on these, let's first take a look at why things go wrong at sundown.

Being up at night: early patterns

A foetus can spend up to 80 per cent of its time asleep, and is more active while asleep than adults are, which explains why expectant mothers can feel kicking round-the-clock. The foetus is exposed to maternal time-of-day cues like rest and activity that synchronise the foetal clock with the external dark–light cycle. However, though day–night rhythms in heart rate can be detected in the foetus, as they are regulated by the mother, they are not running independently.

In Chapter 2 I talked about adults having a wide range of daily body rhythms including peaks and troughs in temperature, blood pressure, hormones and activity levels. Can we detect similar patterns in the newborn? Apparently not; newborns show very little difference in activity levels across the day and night, being only marginally more active during the day. Even at seventeen weeks after birth, babies do not show any marked fluctuations in body temperature. The characteristic human melatonin peak observed around the middle of the night followed by a decrease as the morning approaches is not detectable in babies until they reach the age of two months. None of these findings point to the existence of any circadian rhythm in babies, at least in the first few months. But while there are few behavioural, hormonal or other indicators of a daily rhythm, we do know that the neurons of the body's master clock – the suprachiasmatic nuclei (SCN) – have daily oscillations *in utero* by the end of the first trimester of pregnancy. We also know that the brain pathway activated by light – the retinohypothalamic tract that runs directly from the retina to the suprachiasmatic nuclei – is present at thirty-six weeks after conception, if not earlier. So if the anatomy and some of the functionality is already in place then it seems odd that the pathway may not be responsive. But an explanation could be that cues are mixed because of babies' haphazard exposure to domestic lighting and irregular feeding times. Conditions that are less random, where lighting is timed, presumably let the tract show off any latent potential. Babies born prematurely, cared for under set lighting conditions while at hospital, provide the obvious opportunity for us to find out whether this is the case.

Scott Rivkees and co-workers at the Yale University School of Medicine monitored babies' patterns of activity by a device attached around their ankle, which recorded their movement across the course of twenty-four hours. They looked at activity levels in babies born prematurely for a period of two weeks before discharge from hospital. Babies were either cared for in continuously dim conditions or according to a schedule of light between 7 a.m. and 7 p.m. followed by dimness from 7 p.m. to 7 a.m. Around one week after discharge, those babies who had been exposed to the cycled lighting developed patterns of rest and activity synchronised with the light–dark cycle while those cared for under continuous dim lighting showed a delay in adapting to local light conditions. Although this shows that early exposure to a consistent light–dark cycle is linked with the beginnings of a basic rhythm, as I said earlier, the reality of the home environment and feeding schedule is rather different.

New parents frequently report the most problematic time with a new baby as being around six to eight weeks. Bouts of often inconsolable crying occur particularly in late afternoon and early evening. Babies who show this pattern also tend to have disrupted sleep – persistent bawlers get an average of seventy-seven minutes less sleep than average. Yvonne Harrison from Liverpool John Moores University has made the discovery that babies who sleep well at night have been exposed to more light in the early afternoon. This daily exposure to light may influence the onset of sleep at night by aiding the development of the retinohypothalamic tract. That many child daycare facilities do not have the means to allow babies to get a daily dose of afternoon daylight could exacerbate any difficulties in strengthening the synchronisation of the cycle, while at latitudes north of the Arctic Circle – like the Norwegian counties of Nordland, Troms and Finnmark – the fact that in the winter the sun is continuously below the horizon adds further problems for establishing an activity–rest cycle.

If new parents are already too shattered to face an early-afternoon trip outdoors then another effective intervention is giving baby a bedtime massage. Sari Goldstein-Ferber and co-workers from Tel Aviv University wanted to monitor melatonin production to determine if it suggested the existence of a cycle. As the liver breaks melatonin down into 6-sulpha-toxymelatonin the researchers wanted to find out how much had been excreted during the night, as an indicator of the existence of a circadian rhythm. Each set of parents gave their baby an evening meal and bath followed by a thirty-minute massage. Parents were advised to soothe the baby by touching the infant's head with one hand and lightly stroking the baby's back

93

in a circular motion with the other. This actually turned out to be the least demanding thing parents were asked to do! They were also requested to keep all nappies used by the baby during the night – removing the poo from each one – and separately wrap each nappy in a special bag. Laboratory analysis of the nappies showed that babies receiving this therapy intervention had greater amounts of 6-sulphatoxymelatonin than a comparison group not receiving a nightly massage. The timing of the massage was seen as a strong time signal, reinforcing the adjustment of the circadian rhythm with the light–dark cycle.

But what about some of the other kinds of relationship babies share with time?

Talent 1: foetal exposure to time

From the very beginning, the foetus is immersed in the temporal regularities of life itself. The regular throb-throb-throb of the mother's heartbeat, the pre-mealtime grumblings from around seven metres of gut and the movement of the diaphragm during breathing all create pressure changes and sounds around the foetus. One richly structured source of input is the mother's voice. Because this contains low frequencies, the sound waves penetrate the maternal tissues including the amniotic fluid. Of course, even if the sound registers, this does not necessarily mean that the foetus is active in picking up and processing the information. Although all the essential auditory pipe-work – the outer ear, the auditory canal, the eardrum, the middle ear cavity and the Eustachian tubing stretching from the middle ear to the throat – begin to form early on in the pre-natal period, it is only really around the sixth month of foetal life that the inner ear appears open for business. By this time, the snail-like structure of the inner ear (the cochlea) is much better prepared to deal with the pressure changes that result from the arrival of sounds coming in from the outer and middle ear. Pressure changes in the cochlear fluid set off a whole chain of events in the foetus, like they would in an adult, though some components of the pathway are still immature. An entire community of tiny hair cells launches the transmission of electrical signals to the part of the brain known as the auditory cortex. In adults the exact neural pathways involved depend on the type of sound. If the sound involves complex time patterns like in speech or music different routes are used than for fairly low level background noises like the hum of a fridge.

How can we tell if a foetus extracts anything meaningful from adult voices in terms of information about time? An ingenious experiment by Anthony DeCasper and Melanie Spence from the University of North Carolina showed that a foetus is sensitive to temporal information. They asked mothers to read the same particular passage to their seven-month-old foetuses over the six-week period preceding birth. Then, after birth, DeCasper and Spence played this familiar passage as well as a different, unfamiliar passage to the babies. They found that babies showed increased attentiveness to the familiar piece compared with the unfamiliar one. This showed that the foetus can memorise temporal components of speech like the beat of syllables, the time of onset of consonants as well as the overall temporal order of sounds. It can also discriminate between rhythms. These ensure that the foetus is tuned into learning language by the time it is born.

Talent 2: newborns as speech analysts

When you listen to someone talking, your brain is processing around twenty speech sounds every second. There is now evidence that even babies only a few days old can do this too. A study by Franck Ramus and co-workers at the Institute of Cognitive Neuroscience in London found that French babies could differentiate between Japanese and Dutch as well as between English and Japanese. But with babies hardly being the most articulate of creatures, how did Ramus manage to find this out? He used babies' oral fixation with pacifiers to his advantage. Equipment was rigged up so that each time a newborn sucked its pacifier it would hear a sentence. In this way, after ten sucks, the newborns had clocked up ten sentences in one language. After ten sentences, the situation seemed to take on a 'needle stuck in the groove' feel. Boredom or high familiarity led to a lower sucking rate, which made it the perfect time to introduce the baby to a new language. As soon as there was a switch in language, provided the language was sufficiently rhythmically different, Ramus found that the newborns showed renewed interest and the sucking rate dramatically increased. When languages have similar stress patterns, as is the case with English and Dutch, babies are unable to tell the difference. That babies can discriminate between some languages show that they can tune into time in terms of the rhythm of entire strings of fluent speech as well as its individual syllables.

Talent 3: babies do social time

It will come as no surprise that as adults most of us are highly vigilant in our encounters with strangers. Either subconsciously or consciously we ask ourselves whether this person presents any threat, whether we can trust them, and what they are *like*. But what is not so obvious is that timing plays a central role in enabling such vigilance. We impose a high degree of structure on our exchanges with strangers compared to with those nearest and dearest to us. Cynthia Crown at Xavier University showed that when with strangers we tend to time gazing at or away from each other in coordination in a way that is highly predictable. Couples having relationship difficulties also show this sort of pattern, which suggests that they have regressed to the status of strangers. The high degree of predictability indicates a relationship under distress.

This contrasts with intimates and friends who show a high degree of coordination but who are much harder to predict in terms of *when* they engage in eye contact with each other. Do babies also show different patterns of timing dependent on whether they are interacting with their mother or a stranger? At four months, babies tend to stick to predictable patterns of coordination of the kind used in interaction with their mother, rather than adjusting their timing to match that of a stranger's. Their default mode is thus the reverse of those shown in adulthood with intimate others.

Another kind of social time is the awareness of being in the here and now, where we mentally partition the flow of the world around us into its constituent parts. This concept of a psychological present applies to conversations where the duration of the adult spoken phrase and breath cycle typically lasts somewhere between two and seven seconds. If the conversation runs according to schedule without arguments or being time-pressed, the switches between people will be neat. An adult talking with a stranger will typically take very brief conversational turns – around 2.38 seconds duration, which includes the pause until the next turn. What do babies do? If you have a home movie of a baby at play, you can make your own comparison. First identify an episode around a minute long where the baby is being especially vocally responsive. Take time to familiarise yourself with the interaction's choreography – the sequence of facial expressions (the half-frown, the full facial scrunch, the anticipation, or the lopsided smile), gaze exchanges (blinks, definite looks away) and the spacing of the baby's vocalisations. After this, take a few samples of the length of the baby's turn in order to determine the size of the pause until the adult chips in. When Joseph Jaffe and co-workers in the Communication Science Research Laboratory at New York State Psychiatric Institute looked at timing in babies they found that in the first few months of life the size of the conversational turn is around a second less than the mother's. In terms of being on the path to mirroring adult-like patterns, babies fall slightly short. Why? Daniel Stern from Cornell University says we cannot rule out the possibility that babies of that age are limited by physiology, having a shorter breath rate, or are simply unable to continue their breath cycle through. One alternative explanation is that they are unable to keep track of time. Let's take a look and see.

Talent 4: babies' timekeeping over short intervals

To determine whether babies can keep time we need to design situations that allow us to tap into any proficiency at guessing the duration of short time intervals. John Colombo and Allen Richman from Kansas University sat four-month-olds, one at a time, in front of a screen showing a strictly timed sequence of slides at three-second intervals flipping from light to dark, back to light to dark until, in all, eight slides had shown. When the ninth slide in the sequence was due to show, it was omitted, which of course disrupted the time sequence. How did babies react to this surprise event? Colombo and Richman looked at the precise point of disruption in the sequence and looked for the baby's response. As looking for a change in facial expression would be too subjective, they wanted more reliable evidence. A trick of the trade is to look at changes in a baby's heartbeat. A sign that a baby is mentally engaging with something is when its heart rate dips by around five to twenty beats per minute. Colombo and Richman found that the babies' heart rates did in fact decelerate just at the point when the next display in the series should have appeared. That the timing of the decrease in heart rate coincided with the point of omission of the stimulus – and that this was consistent over more than one time interval – meant that babies could keep time over intervals of short duration.

If babies can register three-second intervals under controlled laboratory conditions, we know that it is not because they are unable to keep time, but that they are unable to match adult duration in exchanges. Instead, what it potentially could mean is that babies do not naturally express timekeeping outside the laboratory. At the time of writing, I am working with Eileen Mansfield and Vicky Lewis at the Open University looking for signs that same-age babies are proactive in timekeeping. Instead of looking at changes in heart rate, we are looking to see if four-month-old babies can actually initiate eye movements to predict events that occur at two- and three-second intervals. As babies can in fact do this, it shows that they are not simply passively responding but adjusting their eye movements to correspond with the size of different time intervals.

Talent 5: babies remembering events in time

In the last chapter, when I talked about experiencing time, this included time expressed in terms of childhood memories, which are inaccessible to most of us. This same level of inaccessibility does not appear to be present if we look at memory recall within infancy itself. Experiments by Carolyn Rovee-Collier and co-workers from the University of Oregon used a fun, simple task to find out if babies could recall an event that had occurred a number of weeks previously. A baby is laid in a crib with a colourful ribbon attached to his or her ankle. Of course, before too long the leg kicks enthusiastically and a note is taken of how many kicks the baby makes. Straight after the number of kicks has been recorded, the ribbon is linked to a mobile. After a period of learning, most babies go into a kicking frenzy. After this, there is a gap of either a week or two before the baby is again placed beneath the mobile – only this time the ribbon linking their ankle and the mobile is missing. The idea is that if the baby recalls the effect on the mobile, then they will resume their kicking frenzy. The number of kicks made at each session is compared to see if the baby shows any sign of remembering previous events involving their own actions. Whereas two-month-olds showed only negligible recall, three-month-old infants were able to recall for up to a week and six-month-olds showed recall over two weeks. From very early on in life, babies have the ability to recall recent experiences and store them over time.

Talent 6: babies' sensitivities to order of events

As adults, we are constantly sensitive to the order of events in time. An everyday example is yanking underwear out of the way and taking aim *before* starting to pee. Jumping to a more complex level of human functioning, a manipulative person will set about a sequence of events in order to find means to satisfy their own ends. First they identify a person's weakness, then they spot opportunities to exploit this knowledge by devising various schemes to induce jealousy, fear or anxiety which place the person on the defensive. This puts the manipulator in the position to achieve their initial aim. Although a number of social abilities allow adults to be like this, a major precursor is the ability to be sensitive to the order of events in time. How early in life can this ability be detected?

In the 1950s pioneer Swiss psychologist Jean Piaget became interested in babies' relationships with time, particularly in terms of how babies ordered events across time. Let's say a baby reaches for an interesting-looking toy. As the baby picks up the toy – a bell – it rings. As the ring depends entirely on the correct ordering of 'pick up' and 'ring', there is a sequential relationship. The event only occurred because the baby's own actions, by chance, led to the sound of the bell. In this way, Piaget only saw babies as capable of sequencing things in order when the natural stream of events led to the conclusion of the bell sounding. As four- to eight-month-olds do not show that they can intentionally link things in order to achieve goals, Piaget believed that it is only around the end of the first year that babies start to consciously coordinate actions towards achieving a goal. However, more recently Peter Willats from the University of Dundee found that nine-month-olds can arrange actions in the correct order to achieve a desired goal. For instance, they can shift a foam barrier to get at a cloth in order to pull an object nearer.

The design of increasingly ingenious experiments has allowed contemporary researchers to credit babies with abilities earlier than Piaget thought possible. Even four-month-olds show signs of ability in serial ordering, provided the task is suitably sensitive. David Lewkowicz from Florida Atlantic University sat babies in front of a screen showing moving icons with some fun sound effects, and showed them this same sequence several times over. Once the babies got bored, they stopped looking at the screen and the trial stopped. Lewkowicz then showed them a new version with the order of presentation switched around and looked to see if this re-engaged their interest. The rationale was that if babies did not detect the event order being switched around, they would continue to behave as if bored. That the babies' interest was renewed indicates that with the right level of stimulation even from an early age babies can note temporal aspects of the world around them being changed.

Talent 7: moving on time

The relationships with time discussed so far have only really emphasised how babies respond to incoming information. But what about turning the tables and looking at the way babies proactively engage with the world around them. Let's look at how they time their actions to the speed of events.

As adults, when we reach for an item on a shelf or go to catch something, in order to manage the action smoothly the grasping movement has to be correctly timed, especially if the object is moving. If the hand closes too early, the object will hit the knuckles; if it closes too late, the object will just bounce off the palm. How soon can such timed adaptations be spotted in babies? There turns out to be a very specific correlation between a baby's age and its ability to time its hand closure before touch. Babies aged five months start closing their hands over 5 milliseconds *after* making contact with an object, whereas nine-month-olds do it around 1.4 milliseconds *before*. Performing best out of all the age groups are the oldest babies, aged thirteen months, who do so around 112 milliseconds before. As this is in line with times adults achieve, it shows that from this age babies are able to time their hand movements in relation to the demands of the environment around them. It is these early abilities in timing which lay the foundations for later achievement in football, tennis, cricket and so on.

Babies do time

With the exception of the rather turbulent start before the circadian rhythm is synchronised with the light–dark cycle, the foetus and newborn baby have sensitivities to time across a range of areas. The six-month-old foetus is immersed in time, and its developing auditory equipment is able to perceive regularities. Once born, it becomes easier to discover babies' sensitivities to time. Despite the amount of time that babies spend asleep, when awake they are highly attuned to information of a temporal nature – be it detecting blips in the order of events, asserting their own timing in baby conversations with others, keeping track of short intervals, the retention of information over time, or the timing of hand movements in anticipation of future events. It is the timing of movement skills that I continue to look at in the next chapter. The millisecond timing success which began in its fullest form at thirteen months comes to fruition in adulthood and is related to sporting success.

Chapter 6

AMONG THE MILLISECOND
TIME ZONE OF MUSCLES

From time to time most of us stub our big toes, drop the odd mug and trip over the cat. Yet even if we are convinced that we are the world's clumsiest person, there is still reason to hold on to hope. Unless we are affected by a neurological impairment, each and every day we carry out hundreds of movements where our actions are coordinated in both time and space. See for yourself; take a quick self-test by turning to the next page.

Admittedly, the act of turning a page hardly requires the same motor skills as completing a back somersault full twist or successfully juggling six balls, but it does require a certain degree of coordination, which many of us probably take for granted. With these actions comes much behind the scenes neural activity carried out over specific timescales. This timed coordination relies on a surprising number of body areas.

Let's meet the main players in this particular feat of coordination. Aside from the obvious ones like eyes, hands, brain and interconnecting pathways, also involved are your forearm, biceps, triceps and deltoid muscles. The rest of your body also contributed to the task by adjusting its posture to compensate while you reached out. We can now look at how your hand and eye timed their work together.

SELF-TEST

Congratulations.
You have successfully
completed the task!

What lies beneath a humble page turn?

Even though you probably were not conscious of this at the time, just before you turned it, your eyes will have glanced at the corner of the page. This glance would have been followed by your fingers curling around the corner of the paper before turning it over. To give you an idea of the kind of timescale involved, in many tasks involving coordination of the hand and eye in space, the hand tends to reach the target location around one second after the eye does. On the surface at least, this sequence sounds like it comprises two steps – a glance and an action – following the intent to turn the page. However, Benjamin Libet from Harvard University has shown that we actually carry out actions before we are consciously aware of having decided to act. Once you decided you were going to turn the page, the brain processes necessary for the page-turning to occur had already begun around half a second before you had any sense that you had made the decision. The relative order of events is not the only dimension involved in turning over a page. The breadth of the hand and arm movement also requires timing. For instance, if we are reading a newspaper, a larger arm movement is needed in order to turn the page than with a pocket dictionary. Surprisingly though, regardless of the dimensions of the reading material, the time taken to turn the page will be virtually the same. This is because the speed at which we flick the page over increases in proportion to the size of the movement demanded by the size of the book.

Now I have described the choreography for the various body parts involved in turning a page, it is time to take a look at what neural activity is responsible for the initiation of arm and hand movement. Irwin Lee and John Assad from Harvard Medical School looked at the timing of arm movements in rhesus monkeys who had been trained to use a joystick to guide a spot of light to a target. In looking for the neural bases for the initiation of this movement, Lee and Assad first homed in at the level of basal ganglia – easily identifiable structures because of their loop-like form. In Figure 6 a cross section shows the three components of the basal ganglia. Next, within the basal ganglia they concentrated on an area called the putamen. Once within the putamen, they continued with this Russian-doll-like approach to locating candidate brain areas and looked in the posterior region, which has overall responsibility for

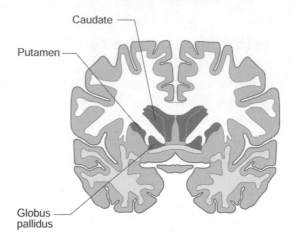

Figure 6 Basal ganglia.

moving the arm. Within this posterior region, they finally scaled their enquiry down to the level of individual neurons. In particular, they wanted to know precisely *when* neurons became excited. Lee and Assad found that neurons tended to fall into one of two categories. Out of a total of 162 neuronal units, they found sixty-nine neurons that were of an 'enthusiastic' sort that fired spontaneously. Typically, these built up to some kind of small but nevertheless noticeable wave of activity hundreds of milliseconds before action started. The second type of neuron, known as 'phasically active', kicked off a second wave of activity, which coincided with the movement of the inner forearm, biceps, triceps and deltoid muscles. In this two-wave pattern, the first wave of activity – a rather low-key event – can be interpreted as paving the way for the actual onset of movement. It is tempting to believe that this pattern of timing could also apply to humans.

Don't keep your eye on the ball

Earlier I mentioned the lag between the eye and the action of the hand in turning a page. A major puzzle is that such a one-second lag would surely make it impossible for many sportspeople to actually play their sport. For a start, tennis players would find it impossible to keep their eye on a ball served at 110 mph at a distance of around twenty-three metres. In cricket, though the sheer weight of the bat and ball are greater, not dissimilar demands are made of batsmen and batswomen where the ball also comes in at a high but unpredictable speed. For instance, faced by a fast bowler he or she only has around 600 milliseconds to hit the ball. They need to look at the right place at the right time in order to determine future trajectory and timing of contact. Michael Land from the University of Sussex and Peter McLeod from the University of Oxford have looked at the timing of eye movement in cricketers to see whether there was anything that differentiated cricket stars from relative novices. They compared the tracking behaviour of three different types of cricketer, professional, amateur and club player, by recording eye movements using an eye-tracking video camera. Balls were launched from a bowling machine at about twenty-five metres per second across a range of different angles. The point of bounce varied between trials so the balls bounced within three and twelve metres of the batsman.

Although McLeod and Land found that each of the types used a broadly similar strategy, the timing of their eye movements depended on skill level. The general sequence of events would start with the cricketer fixating on the bowling machine until the ball's launch. At this point the cricketer would make a predictive eye movement to the location where they anticipated it would land. After the ball had bounced, the player followed its trajectory for around 100–200 milliseconds. Compared to the amateur and club cricketers, the professional was able to more accurately time the shift of fixation from the point of delivery to the predicted point of bounce. One type of shift was rapid eye movements – saccades – which quickly redirect the eyes from one point to another in tracking the ball. The second type was smooth visual pursuit movements, which as the name suggests involve tracking the ball along its trajectory. The least skilled cricketer was much slower in timing his first saccade to the appearance of the ball than the others. Not only that but he also showed much more variability in the timing of saccades. Typically, the two least skilled would tend to wait until the ball had completed a high proportion of its flight path to the point of bounce and therefore spent less time anticipating the movement of the ball.

One of the obvious reasons professional cricketers excel is that their brains have had years of practice coordinating eye with ball and bat. Over the course of our lifetimes, our brains become increasingly adapted to coordinating the primary sense of vision with limb action and the neural architecture becomes used to receiving and coordinating the input with which it is most familiar. Insight into just how neurally attuned to a particular sense we become as we age comes from people who have been blind and then had their sight restored. Researchers Ione Fine and Donald Macleod from Stanford University and the Salk Institute in California met Michael May, who had been blind from the age of three for a total of forty years. Michael was an active skier who negotiated courses by listening to people making commentaries on the slopes. He regained his sight after a stem cell transplant. On returning to the piste after his operation, Michael found that sight gave him a sense of impending collision and he needed to recreate his preoperative state by skiing with his eyes closed. Before the operation, his brain was coordinating sound, touch and sensation. After the operation, and before rehabilitation, despite having access to visual information, his brain remained primarily coordinated in the same triad. What could explain this? What I speculate is that whereas Michael had previously relied on timing of verbal commentary in relation to his own body space, now the predominance of the visual space became directly perceivable, but remained uncoordinated with his body movement. Initially, then, he could not use visual information and preferred to rely on tried and tested associations between the senses he had used in learning to ski. We will revisit the issue of timing and blindness in the next chapter.

Mind the gap: ingoing and outgoing activity

Another method of characterising the nature of the millisecond time zone of muscle activity is to quantify the size of the gap between incoming neural activity and outgoing motor response. Karl Becker and co-workers at the University of Kansas Medical Center looked at an incoming signal being transformed into a motor response by comparing the brain activity of cats across two situations. In the first situation, they looked at cats' neural activity in response to the onset of a clicking noise. In preparation for the second testing situation, they conditioned cats to blink in response to the sound of a click. The beauty of their experiment was the deliberate degree of overlap across these two situations. Although the cat heard the sound of the click in both situations, since the second situation involved an immediate motor response to the sound, Becker was able to distinguish the parts of the neural pathway that attended to the incoming click from those that had duties for the outgoing motor pathway executing the blink. The size of the time interval between the sound and the peak activity of the blink was only eight milliseconds, which shows how fast muscles are capable of responding once they have been conditioned to respond in this way.

One human situation where the gap between sound and response has to be as small as possible is in competitive situations, such as the start of a race. In the case of world-class athletes, the size of the reaction times tells that they are not actually reacting to the starter's shot but actively trying to shorten the reaction – to anticipate without over-anticipating and incurring a false start. Experienced athletes have a representation of the reference time interval virtually engraved on their memories and so can normally estimate the starter's shot with high probability. That senior sprinters are better than junior is reassuring to coaches and athletes because it shows that more accurate representations can be developed through training.

After the starting gun, sprinters immediately rely on a different type of timing. Muscles trained for intense short bursts contain a higher proportion of 'fast-twitch' fibres compared to the composition of muscle fibres of runners participating in endurance events like marathons. In humans, fast-twitch fibres range from 10 to almost 70 per cent of the muscle fibre population, but with a greater proportion found in faster athletes. Compare this with the fastest sprinter on the planet, capable of accelerating to over 60 mph in less than three seconds. Wild cheetahs have as much as 82 per cent of their muscle fibres classed as fast-twitch, which does seem a bit on the low side given the top end of the range for humans, and that cheetahs run three times the speed of an Olympic athlete. Where the cheetah's advantage lies is in the colossal anaerobic capacity of its muscles. This shows that aspiring competitive sprinters not only need to be endowed with a high proportion of fast-twitch muscles but also need to be able to focus their training to maximise efficiency of the major muscle groups at a biochemical level.

What links are there between time perception and movement?

Researcher Ricarda Schubotz and co-workers at the Max Planck Institute of Cognitive Neuroscience have shown that the areas of the brain involved in time perception are also the ones responsible for supporting motor action. It therefore makes sense that the brain houses them together (see Figures 7 and 8). This finding should come as encouraging news for performers like athletes, dancers and musicians or those rehabilitating after injury. If some of the same processes are shared then surely rehearsal without action – like imagining the movements in a particular action sequence – would be an effective aid where normal physical practice was not possible. Support for this idea comes from a study by Daniel Meegan and co-workers at the University of Rochester. In this study, people listened to two tones in succession, differing in duration, only one of which remained the same length across different trials. The idea was that after 2500 trials people could be expected to have built up a robust representation of the standard duration at a neural level. People were then tested on a separate task to assess their accuracy in physically reproducing the target duration, and their performance was compared to the baseline level before the trials took place. The researchers found a large improvement on the trained duration relative to non-trained durations, which showed that sensory representations directly impact on motor representations. Being exposed to temporal cues presented in one mode has the potential to transfer without necessarily having to go through the physical motion during learning.

Moving to the beat

So far I have mainly talked about how we adapt our responses to unpredictable events. But what about the ways sportsmen and -women synchronise actions when the pace is predictable? Events like rowing and rhythmic floor gymnastics demand that competitors stay locked to a predictable external beat. Our ability to synchronise movement with external pace has been investigated under stringent laboratory conditions. Typically, researchers monitor our

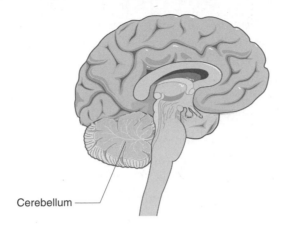

Cerebellum

Figure 7 Cerebellum.

aptitude at tapping in to synchrony with external schedules. If you have a metronome, you may want to try this out. Set the click at a manageable rate of say 0.8 seconds and tap your finger in time to the click. If you're like most people, you will judge your taps to be in time with the pace of the metronome. Surprise, surprise – the reality is somewhat different. We tend to tap between about twenty to fifty milliseconds ahead of each click of the metronome.

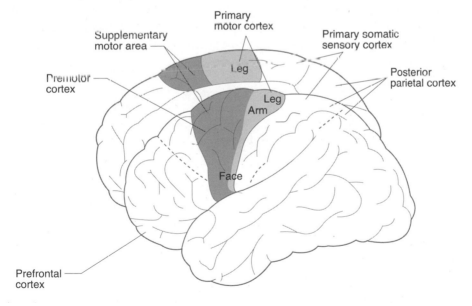

Figure 8 Location of brain's motor architecture.

Why? One explanation for jumping the gun is due to the sounds needing to be synchronised internally. Another explanation is that the brain's processing of the sound of the click of the metronome occurs at a different rate to the processing of feedback about the timing of the tap.

However, when Bettina Pollok and co-workers at the Max Planck Institute tried to find evidence of brain activity commensurate with these findings of differential processing rates, the outcome was disappointing. Pollok found that when people tapped, the main area responsible for processing this activity was the primary sensorimotor cortex. The motor command for the tap caused one ripple of brain activity. The feedback from delivering the tap caused a second wave in the same area. Although the research team detected a third wave of activity, it was not a large enough effect to provide convincing evidence that an actual comparison was being made between the timing of the movement of the taps and the click of the metronome. On the basis of this trial, it is still early days in identifying the brain activities relevant to this task. Perhaps part of the reason for the difficulty in terms of identifying brain activities is that the neural pathways involved differ according to what medium the pacer is presented in. Researchers from the Otto-von-Guericke University of Magdeburg found that the timing of taps was superior on auditory tasks, was accompanied by more activity in the brain and took different neural pathways, than on vision-related tasks. Based on these findings, it is tempting to speculate that the domination of the foetal experience by sound-based information (Chapter 5) primes our neural pathways for life.

Improving timing skills in sport

Talented sportsmen and -women routinely rely on the use of imagery in training, rehabilitation from injury or immediately preceding action. In a run-up to a jump, prior to a penalty kick, sprint or throw, they mentally move through the actions, visualising a successful outcome. There is strong evidence that athletes who visualise movements in terms of timing, rhythm, direction and amplitude perform at a higher level than those who do not. One factor affecting the efficacy of mental rehearsal is the pace of playback. Claire Calmels and Jean Fournier at the National Institute of Sport and

Physical Education in Paris looked at twelve top artistic gymnasts. Each gymnast completed two tasks – first, the physical performance of the same competition-ready floor routine and second, imagining themselves performing the routine in competition. During this second task, the gymnasts were asked to signal the different specific stages of the routine by giving taps. The routine in mental imagery was found to average sixty-five seconds, compared to the length of the actual routine, which averaged eighty-three seconds. Where did eighteen seconds get dropped? Calmels and Fournier discovered from the taps that one particular stage in the routine appeared to be shorter than all the others. It appeared that the gymnasts found this part of the routine somewhat easier to execute than the other parts. In contrast, the part of the routine judged to be most difficult by the International Gymnastics Federation actually took longer to run through mentally. These findings imply that the use of imagery may be beneficial for young athletes to encourage them to slow down during early acquisition of a new skill. As the difficult sequences take longer to represent mentally, the key is to run through them more slowly at first.

Time zones of muscles depend on the nature of imposed demands

In this chapter I have talked about the millisecond time zones of muscles in terms of the relationship between the conscious decision to act and the action, the speed of overall movement, and the phased timing of neural activity. In certain dynamic, unpredictable sports the timing and distribution of eye movements are diagnostic of the level of skill that has been achieved. Training in the most effective use of eye movements can clearly make a difference between a hit and a miss. Any discussion of the time zones of muscles must consider the nature of the demands of the environment – like whether the task is predictable and whether it is presented via sound or sight. The outcome (that people perform to a higher standard on predictive tasks which involve sound rather than vision) falls in line with the finding I reported in Chapter 4: that the visual clock runs more slowly than the auditory clock. From this it is tempting to speculate that the domination of the foetal experience by sound-based information – discussed in Chapter 5 – biases our neural pathways for life.

Of course, not all of us have muscles that perform at the same kind of speed as those of sports stars or even the majority of the population, and the next chapter takes this into account by exploring our more exceptional relationships with time. These unusual relationships are not reflected purely in terms of speed of muscle movement, but highlight various conditions affected by time – addiction, sensory impairment and autism to name but a few. As you will see, in some conditions just one relationship with time has been affected, whereas in others a number are affected.

Chapter 7

DIVERSITY IN RELATING TO TIME

In an increasingly automated world our pace of life is determined by the preprogrammed features of our environments. Car-park barriers, lift doors, ski lifts and automatic sliding doors all require us to complete manoeuvres before a cut-off point. In East Genesee Street, Skaneateles, New York State the pedestrian crossing has a twenty-four-second countdown timer. In Tianjin, China traffic lights display a red or green signal that shrinks as the time runs out. As for the phone, in the UK you only get five seconds to dial before being told that the number has not been recognised (in the USA the delay allowed is very much shorter). If you dial the right number but no one answers, you get nine seconds to collect your thoughts and prepare your message on a standard answerphone before the beep. Commuters trying to buy a ticket who do not manage to select the right sequence of buttons followed by the correct insertion of coins within the allocated time period find the whole transaction cancelled. If you have a movement disorder or are slower due to fatigue or ageing, then you experience numerous aborted attempts at all sorts of action.

Earlier on, in the first three chapters, I touched on diversity in people's relationships with time when I compared pace of life, access to the twenty-four-hour economy, longevity and organisation of life. In this chapter I revisit the issue of diversity, but in terms of how certain human conditions and individual characteristics affect people's relationship with time. By doing this I provide opportunities to understand more about how people differ from each other and the potential ways we can adapt our living spaces or approaches to work. First, I look at some selective examples of movement disorder and what specific difficulties these cause.

115

Brain dysfunction and timing movement

In Chapter 6 I named three brain regions with housekeeping responsibilities for time perception and motor action – the cerebellum, basal ganglia and the premotor cortex. It is hardly surprising that if dysfunction occurs in any of these regions, people's timing will be affected in some way. Let's now take a look at a selection of disorders of movement associated with brain dysfunction in these areas, and find out whether any abilities are spared.

To laugh, shriek, shout, cough, whisper, sing or cry is virtually effortless for most of us, yet a highly coordinated sequence of muscular activity is needed in order to produce these sounds. For instance, even a humble cough requires the timed cooperation of seven muscles in the larynx. An award-winning study by Allen Hillel from the University of Washington School of Medicine looked at how these seven muscles work together. Hillel's study involved threading electrode wires through fine needles and inserting them into volunteers' muscles to record timings. Apparently only one person dropped out, which is surprising given that no one participating in the study was sedated when the needles were inserted. The findings showed the timetable of activity during the elapsed time window of 1.2 seconds. Three adductor muscles contract over the first 300 milliseconds. Then around halfway through this, the abductor muscles start their shift. Then, for about 150 milliseconds, the entire team of muscles works together, contracting in unison. Then the adductors abruptly stop their duty while the abductors get set to springload the glottis in order for the vocal cords to open. Finally, as soon as the abductor activity stops, the adductor muscles contract over the last 600 milliseconds.

One reason Hillel was keen to unpack what movements lay behind a cough as well as other kinds of sounds was so he could find out more about the timing of movements in people affected by dystonia of the larynx in comparison with those with no difficulties. People with dystonia have a basal ganglia dysfunction and experience involuntary muscle contractions that cause repetitive twisting, odd posture, muscle overactivity and fatigue. Any number of body areas can be affected including the neck, eyelids, face, arms, legs, trunk or vocal cords. In the case of dystonia of the larynx, the most common symptoms include difficulty initiating vowel sounds, an abnormal pitch of voice and a tight sensation when talking. When Hillel looked at the performance of individual muscles he found that timing was dependent on the type of activity being performed. Functions like coughing, swallowing, throat clearing and respiration were all spared from showing any unusual profile, because none of them involved the voice. When Hillel looked at subjects' performance on speech tasks (for example saying 'We mow our lawn all year') a dramatically different picture emerged. Whilst unaffected people can start to say this some 233 milliseconds after the onset of activity in the TA muscle, in comparison, people with dystonia are substantially slower off the blocks. Not only that: while in unaffected people the TA and the LCA muscles rest for 90 per cent of the sentence, in those with dystonia the muscles work overtime. These findings show that even timing difficulties occurring at the scale of milliseconds can affect the ability to communicate.

Another condition caused by dysfunction of the basal ganglia is Parkinson's disease (PD). A hallmark of PD is an overall slowing of movement thought to be due to a deficiency of the neurotransmitter dopamine. In Chapter 6, I mentioned that there were commonalities between time perception and action. Given the slowness of action, it would be plausible to suspect that people with PD would also tend to overestimate the duration of time at a mental level. In line with this, Chara Malapani and co-workers from Colombia University found that when tested with intervals in the twenty-second range, people with PD would indeed tend to overestimate. But the story is a far from straightforward one. When patients were presented with more than one time interval in a single sitting, they tended to overestimate the size of a short interval (for example eight seconds) but *underestimate* the duration of a long one (for example twenty-one seconds). One possibility is that the creation of the memory for the first task interferes with the recall of the second. Whether or not this occurs on the way to memory storage or at the retrieval stage has yet to be discovered. In any case, the picture is further complicated by the discovery that there are situations where the characteristic slowness of movement of PD is lessened. Researchers have repeatedly found that when a cue, such as a sound, is provided in advance of the time that the movement is to take place, the movements are less impaired. Other neural structures take over the reins from the basal ganglia and timing abilities are spared. Under laboratory conditions people affected with PD are able to achieve a goal in tune with the pace of the environment around them.

117

Another brain area with responsibility for timing is the cerebellum, to the extent that if someone has damage to this area (for example due to a lesion or a malfunctioning thyroid) an array of problems with timing follows. This could include things like difficulty tapping in time, sequencing movements, making accurate judgement of the duration of tones and determining the velocity of moving objects. Jon Hore and co-workers from the University of Western Ontario found that people with lesions in the cerebellum have difficulty timing an overarm throw. Hore and his team looked at the timing of finger opening to release a ball. They compared recreational players with skilled throwers who were competitive cricket or baseball players. Dependent on skill level, there was a difference in when the launch took place. The researchers timed the moment from when the ball is released from the fingertips to the moment in the throw when the hand is vertical in space. Whereas recreational players took on average ten milliseconds, the best players only took two to three milliseconds, while people with cerebellar lesions took fifty-one milliseconds. Given the large differences between the groups, Hore identified the problem as being to do with the actual release of the ball, but questioned whether the problem was isolated to being unable to time the fist opening correctly or whether it was due to the coordination of timing across body parts. For example, if it was a matter of co-ordinating information from two different body parts, this would tell us that the cerebellum has some kind of overseeing role in combining the motions of different body parts into a coordinated movement. Although it is early days in this field of research, the process of identifying the specific difficulties experienced by this group will allow the essential components of a successful throw to be isolated, which will in turn have implications for informing the coaching of bowling.

So far I have talked about disorders in relation to two separate areas of the nervous system. But what about conditions that appear to be due to malfunctioning in *both* of these areas?

Timing and attention-deficit hyperactivity disorder

Children diagnosed with attention-deficit hyperactivity disorder (ADHD) are observed to be overactive, impulsive, inattentive and to show difficulties processing language. One explanation for this condition is that these children have trouble concentrating and this disrupts their timing abilities. There is some support for this theory, as children tend to underestimate durations, particularly those in the range of twelve to sixty seconds. Anna Smith and co-workers at the Institute of Psychiatry monitored affected children's ability to handle intervals matching the size of the units in everyday speech to see if they could explain some of their difficulties processing language. They also compared ADHD children with matched unaffected children selected for having comparable skills at keeping track of information, like recalling a phone number. While the ADHD children generally performed in line with controls on the longer time intervals, they turned out to need fifty milliseconds longer than the controls to discriminate the duration of two stimuli. Although this length of time may seem so small as to be insignificant, as I mentioned in Chapter 5, twenty milliseconds represents the duration of the smallest units – called phonemes – of speech, so this may explain some of this group's difficulties in processing language. Interestingly, Becky Shaw and co-workers at the Open University have found that when playing computer games, children with ADHD can anticipate time intervals and coordinate timed motor responses over short periods of time. If there is an adaptation in the form of a structured display, with repetitive components providing the necessary support for them to remain motivated, then these children can perform as well as control children on these games.

Experience, memory and time

In Chapter 4, I talked about our capacity to project thoughts both backwards and forwards in time. One aspect of this ability involves the need to remember daily life tasks like appointments. People affected with Alzheimer's disease find this kind of memory task increasingly difficult as the disease progresses. 'Tangles' and 'plaques' develop in the structure of the brain, leading to the death of cells. As a consequence, things like remembering to take medication, paying bills, dressing and going out become increasingly troublesome without the support of a carer. Another relationship with time problematic for this group is the disruption to the sleep–wake cycle. Damage to various neuronal pathways responsible for initiating and maintaining sleep causes night-time disturbance, which is often cited as the reason a family with a member affected by Alzheimer's takes the decision to seek alternative care. In the UK the decision to admit someone with Alzheimer's to a specialised facility increases the total care costs to the state by an average of £20,688 per year, so any interventions that serve to slow the transition to professional care reduce the burden on the public purse.

Help for Alzheimer's sufferers, whether pharmacological or otherwise, is critical because the disease currently affects 19 per cent of people aged between seventy-five and eighty-four, and 47 per cent of those over eighty-four. Given that the UK has an ageing population, the numbers affected by Alzheimer's will rise, and the need to develop cost-effective ways to delay impairment will become more pressing. Pilot work has looked at the potential for the use of electronic memory aids and early findings are looking promising for those mildly to moderately affected. Such devices allow the vocal recording of specific appointments, which can then be preprogrammed by date and time. An alarm sounds at the required time, providing a verbal reminder of the appointment.

It is not only people thought to have Alzheimer's disease that have more than one altered relationship with time. In Chapter 2, I mentioned people who have had their eyes removed and who are consequently unable to synchronise with the local light–dark cycle. And in Chapter 6, I mentioned how Michael May continued to rely on time-based information to determine his movements in space. Now we'll take a more in-depth look at the relationship between sight and time.

Sensory impairment and time

That people who have had both eyes surgically removed due to tumours do not synchronise with the local light–dark cycle does not necessarily mean that people visually impaired due to other causes will also experience poor sleep quality. Damien Leger and co-workers from the Centre du Sommeil, Paris, looked at the sleep quality of twenty-six adults who had no perception of light and were blind. They found that regardless of whether the cause was congenital blindness, acquired blindness or having prosthetic eyes, the total sleep time, onset of sleep and quality of sleep for these subjects were all lower than for the comparison group. Their bodies' melatonin production occurred uniformly around the clock rather than being suppressed during the day by the presence of light. As a result, their rhythms were 'free running', even though all had the structure of working life and the social cues of their family. This outcome shows how much the biological clock depends on access to light stimulation via the eyes – despite the existence of clear social cues, the circadian cycle is disrupted.

In Chapter 5, we saw that babies are naturally exposed to the pace of talk, smiles and eye movements of those who care for them. But what about babies born deaf and blind, unable to learn the pace of life through watching or hearing the world around them? Researchers have found that babies born blind and deaf make briefer movements than children with no sensory impairment. Their actions are short with more repetitive qualities like rocking, eye-poking and so on. The lack of sound and vision may be leading them to gain more stimulation by self-initiated movements, which are necessarily shorter because the babies are motivated to repeatedly seek out stimulation from outside themselves.

While sighted babies enjoy lots of spontaneous face-to-face eye contact, babies blind from birth and their parents rely on communication expressed through sound and touch, which is much more sequential in nature than visual communication. Because of this dependence on sound and touch, the play between child and parent relies on the properties of these two senses rather than the here and now of the visual world. This experience and the sense that time is a source of information critical to people who are blind comes from the reflections of people like John Hull. In his autobiography *Touching the Rock* he says:

> This sense of being in a place is less pronounced . . . Space is reduced to one's own body, and the position of the body is known not by what objects have been passed but by how long it has been in motion. Position is thus measured by time . . . For the blind, people are not there unless they speak . . . People are in motion, they are temporal, they come and they go. They come out of nothing; they disappear.

As speech sounds naturally occur in a stream, they come as readymade time-structured activities highly suitable for visually impaired babies and their carers to capitalise on. My own research with blind babies and their parents has shown that the predominant forms of play are highly structured routines repeated over and over – more so than for sighted babies at a similar language level and matched for a whole host of other family factors. In one family their congenitally blind son of fifteen months would touch a surface with a variety of textures – like sandpaper, or cellophane stuck onto pieces of card. As his hands traversed the textural landscape, his mother would vocalise a description – 'Very rough . . . very rough . . . oh and onto the crinkly paper' – to coincide with the changing material, sound, texture and function of the objects he was exploring. Having a sequence of labelled events presented at carefully timed points is important for children who are blind as it enables them to make a link between toys and the words we use to label them. Not having the same joint visual space as sighted people to operate in, the links in time between experiencing objects and having them described become crucial.

Learning to identify a toy by touch alone represents part of the challenge a child who is blind or severely visually impaired faces in learning to talk. Once I overheard a blind three-year-old chatting with his mother over lunch. 'Mum,' he said, 'what's the time?' His mum replied, 'Ten past twelve.' He immediately asked, 'No, what's the time now?' His mother repeated, 'Ten past twelve.' Using a more determined voice he persisted, 'No, what's the time *now*.' She again repeated, 'Ten past twelve,' and he came back with a final round of, 'No . . . the time *now*.' The exchange showed that this three-year-old had a sophisticated awareness of the concept of time being an ongoing passage without visual access to watch or clock faces.

Autism and time

In Mark Haddon's award-winning book *The Curious Incident of the Dog in the Night-Time* the role of routines, repetition and rituals in the day-to-day life of Christopher Boone – a fifteen-year-old with Asperger's syndrome, a form of autism – was paramount. Although autism exists along a continuum of severity, people with autism generally spend a good deal of time being concerned with the replication of events, which suggests that they operate largely in the present and in the short-term future. People with autism have little sense of their own lifespans – they do not engage in planning their future – except, as I explain later, some individuals do have remarkable abilities in relation to future time.

Autism is diagnosed when people show impairments in social abilities and communication skills, and when behaviours and interests become increasingly restrictive and repetitive. In their social play autistic children show minimal interest in other people and are unlikely to share any of their own interests. They are also unresponsive to other people's emotions. Of the 50 per cent of children with autism who acquire any useful language, their speech may have an unusual tone, the timing of their turns in conversation may be poor, and they may interrupt. They also tend to repeat words, phrases, jingles and other familiar sentences. They have particular difficulty with language that is not literal, so irony, sarcasm and metaphor are neither used nor understood. At certain times, some people with autism show behaviours that are quite stereotyped – repeated rocking, flapping hands or even self-harm. Autistics may obsessively arrange objects in a particular way or display a deep fascination with collections, order and detail. Depending on their age, this may be expressed through strictly ordered daily routines rather than through the acquisition of objects and materials.

To return to conversational turn-taking and other aspects of social timing, people with autism often display an inability to take part in the flow of social exchange. Their timing is quite literally off: they find it hard to orient their attention to a new cue in the environment until some 800–1200 milliseconds after the event. This compares to unaffected people, who manage to respond in under 100 milliseconds. Such difficulties with switching attention have been linked with a lack of Purkinje cells in the cerebellum. Observations of mice with Purkinje cell loss have shown that they too show a cluster of difficulties with timing, like hyperactivity and repetitive behaviour. These findings hint at one possible neural basis for autism, but it is still too early for answers.

Autistics also show disturbances in their circadian rhythms. Although melatonin levels are in phase, being suppressed by day and secreted by night, the size of the change is much smaller between night and day, which may explain why children with autism have sleep problems. Against a comparison group, parents of autistics reported that their children had more instances of night-waking and difficulty settling to sleep.

Calendrical calculators

Around 10 per cent of people with autism are 'high-functioning', in that they are of average intelligence, use language and can show an exceptional ability. Often called savants, they have areas of outstanding talent in drawing, music or arithmetic skills, including calendar calculating, the ability to name the correct day of the week for any given date. Special talent in calendar calculating is more common than in drawing and music, and emerges between the ages of eight and fifteen, though it has been seen in children as young as six. Some calendar calculators give answers within a couple of seconds, but all are most accurate for dates in the very recent past and slow in their responses the greater the time gap from the present day. Let's look at what lies behind this ability.

One defining characteristic of calendars is that they are highly structured and repetitive. For example, in the Gregorian calendar, the same date falls on the same day of the week every twenty-eight years. This regularity of structure may resonate with one of the hallmark characteristics of autism, the adherence to order and the tendency to focus on patterns and structure, and autistics' devotion to repetition and practice has been lined up as the most likely explanation of this phenomenal talent. Linda Pring and Beate Hermelin at the University of London found that a savant called Peter could not only provide, within four seconds, the name of a day of the week for a date falling in the previous six decades but he could also excel at an unrehearsed task. When tested on various letter and number associations where the tester called out a number and Peter gave the name of the letter of the alphabet corresponding to the number (for example A/1 and Z/26), he performed as fast as a professor of mathematics, no matter where the letter was in the alphabet. A particularly outstanding feature was Peter's lack of errors. His performance shows that a key characteristic of calendrical calculators is the capacity to spontaneously recognise new rules and relationships.

A case for impaired relationships with time?

Based on the various difficulties with time experienced by children with autism, Jill Boucher from the University of Warwick has suggested that people with autism have deficits in relating to it. The lack of full 'entrainment' with the local light–dark cycle as well as the lack of social rhythms creates a separation from other people and life.

What would be revealing would be the chance to be able to track the circadian rhythms of children right back from the pre-natal period. As autism is currently not typically diagnosed until the age of two or three, this remains an important topic for research.

Addiction and time

Up to this point, I have looked at various conditions which arise from brain dysfunction or some other impairment. Now I switch to looking at self-inflicted conditions and how these affect relationships with time. Lapp and co-workers from the New York Research Institute on Addictions discovered what most of us know intuitively: that time appears to pass more quickly when you're drinking. They asked social drinkers to rate their subjective experience of time before drinking, at peak blood alcohol concentration, and post-peak blood alcohol concentration. Volunteers received an alcohol-free drink or alternatively a low dose or high dose of alcohol and were given information about what they had drunk in line or at odds with the truth. For regular drinkers, alcohol may help to increase the subjective flow of time – the illusion that time passes more quickly – as events appear to speed up. This speeding-up effect also occurred even when people thought they had drunk alcohol when in fact they had not. It seems that it is enough to *think* that you're drinking to believe that time is passing more quickly, which means that feeling relaxed may also affect our perception of time passing.

Does exposure to alcohol in the pre-natal period cause timing difficulties in the next generation? Tara Wass and co-workers from the University of Tennessee looked at timing behaviour in fourteen children identified as having a history of heavy pre-natal alcohol exposure compared with children without such a history. The subjects watched a visual display with a series of runway-style lights, and were asked to indicate when a target reached the end of the display. The children with alcohol exposure made more errors and were less able to accurately reproduce timed movements within target times, suggesting some kind of damage to the central nervous system – quite probably the cerebellum and the basal ganglia.

People with a brain disorder due to chronic and excessive alcohol abuse – known as Korsakoff's syndrome – have a diminished sense of time. Matthias Brand and co-workers at the University of Bielefield found quite bizarre responses to tests asking forty-one patients with Korsakoff's syndrome to estimate the duration of everyday activities. For instance, when asked how long it takes to have a morning shower responses ranged from fifteen seconds to one hour. Similar responses were noted on other tests concerning duration of events. One possible explanation is that in such cases the internal clock is deactivated. Another is that the ability to recall memories is compromised. As the underlying damage due to ethanol is widespread it is difficult to pin down which area is responsible for which loss, and in any case there are likely to be a number of shared brain areas responsible for the deterioration.

Another way the relationship between addiction and time has been explored is through the concept of 'time perspective'. This stems from the work of psychologist Philip Zimbardo at Stanford University. Zimbardo administered questionnaires to hundreds of people in the USA to see if there were differences in the way they thought and acted in relation to time. On the basis of his work Zimbardo identified three different mindsets – past-, present- and future-orientated. Someone with a mindset focused in the present makes decisions in the here and now, using observable aspects of the situation rather than relying on future consequences or information from the past. They might routinely walk into a shop and spot some eye-catching deal – maybe a buy-now-pay-later offer on a sixty-inch flat-panel TV – and prioritise this in their memory ('Well you only live once') rather than consider their nagging mortgage arrears. Researchers split people who function in a present time orientation into two further subgroups – 'self-indulgent pleasure seekers' living in the hedonistic present ('I feel that it is more important to enjoy what you are doing than to get the work done on time') and fatalists ('My life is controlled more by my destiny than by my actions'). People who are past-oriented rely predominantly on the currency of reconstructions of previous events ('I enjoy stories about how things used to be in the good old days'). Living in a past orientation means playing it safe and keeping a sense of stability. For a person who is strongly future-oriented ('I believe that a person's day should be planned ahead each morning') it is all about achievement. These people invest in the future consequences of their actions and strive for long-term gratification, managing to avoid distractions and temptations. People in this mode aspire to become increasingly efficient and in extreme cases do not want to 'waste' time relating to family or friends. Various supposedly positive outcomes associated with a future orientation include fewer risk-taking behaviours, higher socio-economic status and superior academic performance. Each of these time perspectives provide a measure of a person's individual orientation in time, and depending on the position adopted, this will influence the kinds of judgements, decisions and actions taken.

A survey by Keough, Zimbardo and Boyd showed that those who lived in the present were apparently more likely to use alcohol, smoke and use drugs. Having a high mental affinity towards the present is also said to be a greater predictor of risky driving, gambling, mental health problems and crime. According to these researchers, people living in the present do not consider outcomes like future poor health, loss of a driving licence, accident, fatality, debt or loss of friends when they act.

Insights into how drug addicts relate to time can be gained by looking at whether they tend to live with a particular time horizon. Nancy Petry and co-workers from the University of Vermont looked at the time perspectives of thirty-four heroin addicts who had on average a 9.4-year history of addiction and had received three months' treatment. Petry found that addicts scored lower on future orientation than a comparison group and correspondingly higher on the present hedonistic and present fatalistic scales. Petry also looked at the ways in which they thought about their future. Addicts were asked to list ten events that might happen during their lives along with the expected age at which each event might be expected to occur. If you want to see how you compare to Petry's group, choose ten events. Then for each event work out how many years in the future you expect it to happen. Add the numbers together, divide by ten and you're now ready to compare Petry's findings with yours. Based on people aged between eighteen and fifty-six, the heroin addicts predicted ten events would occur on average 5.4 years in the future whereas the equivalent for the controls was 8.8 years. Another difference was that whereas the addicts projected an average of eighteen years into the future, for the control participants it was twenty-six. These findings show that addicts live in the present or have shortened perceived futures. This is possibly because they have accepted their addiction and know it has probably shortened their lifespan.

In another task, addicts listened to a number of stories each containing brief time references. For example, 'After awakening, Bill began to think about his future. In general he expected to . . .' They were asked to finish the sentence and provide a timescale. Whereas the average for the group addicted to heroin was nine days, the average for the comparison group was 4.7 years. There is the possibility that coaching addicts to develop a future time perspective may serve as part of their treatment. It also follows that encouraging children to take a future perspective may help prevent substance abuse. There is evidence that the onset of tobacco, alcohol and marijuana use in eleven- to twelve-year-olds is related to having more of a present orientation, but the influence of peers may ultimately remain more powerful, at least in getting started.

Patterns in diversity

By looking at various impairments, conditions and lifestyles it is possible to see a range of relationships with time in terms of muscle timing, time perception, experience of time and entrainment with the light–dark cycle. People with a visual impairment, autism or Alzheimer's appear to have several relationships with time affected by their condition, while those with a brain dysfunction linked to a movement disorder tend to be compromised in one sort of relationship. In this quick glimpse of some selective conditions I have also showed how autistic people with levels of high functioning show exceptional relationships with time. Unusual behaviours with time need not be seen as deficits, but rather as contributing to diversity in terms of the way people relate to time.

Although this chapter is intended to illustrate how people with various conditions relate to time, it also serves as a foundation from which to understand the ways people in general relate to it. Understanding this diversity will mean that we should be able to inform the design of time-related aspects of our surroundings. Just as there are scissors for left-handed people, it should be possible to adapt gadgets, interiors and lighting to reflect individual relationships with time. This could optimise the ways the environment suits our limb-pace (e.g. dialling speed) and rest–activity preferences and individual time horizon perspectives. In future, we may come to know our finger speed in the same way we now do our shoe size, and be able to select a dialling speed or anticipated length of message on an answering machine. And we may be able to customise our homes so as to program a certain lighting intensity dependent on the time of day and season, so that we have the best environmental support to our health.

Chapter 8

BEYOND NINE TO FIVE

Up to now, each relationship with time has been kept boxed up in its own chapter, with recurrent themes across chapters only receiving mentions in passing. Now, as the book draws to a close, the different threads of our relationships with time are woven together, offering a vantage point from which to view the lived time in which each of us is immersed, as well as to provide a forum in which to remind ourselves of the world's temporal inequalities, and to focus on thinking of time beyond the nine to five.

Locating yourself in time

Even while you are doing something as ordinary as reading this page, a host of the various relationships you hold with time are in full swing. Aside from obvious conventional relationships, like being able to locate yourself according to Greenwich Mean Time (if absolutely necessary to nanoseconds) or public time according to either the Gregorian, Jewish, French Revolution or some other calendrical system, at a micro level you can locate any instant in your life according to various cyclical events of bio-time. About once every seven seconds you inhale. Around once every second, your heart beats. At a neurological level, there are two timekeeping systems, a master circadian clock located in the suprachiasmatic nucleus (see Figure 2) which is synchronised to the light–dark cycle, and an internal timing clock that keeps track of the passing of time while you are reading. Around every twenty-four hours your body clock starts a new rhythm, and so, depending on when you are reading this, you will be passing through a particular point in your circadian rhythm.

If you happen to be having a sleepless night and are attempting to read yourself to sleep at 4 a.m. – the trough in the rhythm – then your fine motor skills will be compromised, resulting in page turns being more fumbled than if you were reading this during the hours of daylight. If you are reading this outdoors in broad daylight, the page could be illuminated by something verging on 100,000 lux, ensuring that the light entering your eye continues the suppression of melatonin, so regulating sleep in relation to the duration of darkness. And this is not the only case where the eye plays a role in regulating the quality of your relationship with time. At this moment your eye muscles (or in the case of Braille, your finger muscles) orchestrated by your brain are making a series of timed movements from left to right across the print, reminding us that humans live and act in time's arrow, in linear progression. While reading this book, let's say your eye or finger muscles are not the only active ones. Instead, you are relating to time polychronically by reading while sitting on an exercise bike and timing each push on the pedals to the beat of music. According to research, possibly as a result of the brain's different processing rates of the feedback of the rhythm and the action, you are jumping the gun, and likely completing each pedal revolution slightly ahead of the beat.

If your eye muscles and speed of processing information have allowed you to achieve an average reading speed, then since starting this book you have lived through around 16,318 heartbeats, amounting to some 0.0007 per cent of your life, at least according to an average world life expectancy of sixty-five years. However, the equivalent statistic if you were a Japanese citizen would shrink to 0.0005 per cent, compared to if you lived in Botswana in 2010, where it would balloon to as much as 0.002 per cent. The habits you adopt while reading may also further influence the size of the percentages. Let's say you've been religiously sipping green tea throughout reading this book and also happen to live in a country largely free of everyday threats to survival – political unrest, malnutrition, dehydration, inadequate shelter, frequent risk of infectious diseases, pollutants and inadequate access to medical care. The seemingly insignificant act of sipping such an antioxidant-laden drink would, if repeated over time and in conjunction with other pro-longevity habits (while obviously avoiding the bad ones), serve to increase your probable lifespan.

As you read this, you are also locatable in terms of generational time.

Although three-generation families are commonplace today, the relatively recent phenomenon of the increased amount of spacing between generations directly influences another relationship with time: how people use it. In Britain, around one in ten women spend a high proportion of their time caring for their parents while also still having a dependant in the household. Meanwhile the division of limited leisure time will inevitably fragment down time into crumbs, which arguably limits the capacity of the body to recuperate and to provide optimal protective benefits to the immune system.

Another way to locate yourself in relation to generational time is to look at the impact you are having on any unborn generation's relationship with time, and possibly your own, too. First, regardless of whether you are male or female, there is the possibility that late childbearing will increase your own longevity, thought to be due to the protective powers of oestrogen in the case of females and the positive social benefits of parenting in both cases. Second (and on the grim side) certain DNA strand and protein damage caused by pesticides, tobacco smoke and air pollution affects sperm and hence the survival of future generations and may cause birth defects or longer-term problems such as cancer, chronic disease or infertility. In fact it is now known that pesticides and other man-made chemicals have the potential to damage sperm or lower sperm counts across as many as four generations. Another significant influence on future relationships with time is pre-natal exposure to alcohol, known to cause brain damage and compromise mental functioning. This can have a negative impact on a child's ability to perceive time, a capability known to be crucial for skilled movement and successful social interaction. Third, and on a slightly more positive note, your body will be providing a continuous stream of both regular and unpredictable sounds to your unborn child, which will prime its auditory system to discriminate the phonemes of your shared native language, and will also attune it to the rhythms of the human voice. It is these early experiences, which induct us into the temporalities of our social and cultural worlds, that we continue to assimilate when we start to learn to talk. As an adult, if someone were to interrupt your reading of this sentence, you would automatically process their speech, discriminating the tempo and rhythm of the sounds in order to find out what they wanted. You would also mix in a series of timed vocal turns and eye gazes, which, if you were communicating with a stranger, would be matched to theirs more closely in terms of duration than to those of someone close to you with whom you normally enjoy a harmonious relationship.

Whether you are reading this in leisure time, work time or some blurred space between the two will of course depend partially on the reason why you began reading this book in the first place. Let's say you are working in a clock time culture, reading this under pressure in order to produce a review in time for a 5 p.m. deadline that you now wish you had never agreed to. If this is the case, then as the deadline approaches you will be calling on the services of your internal timekeeper while using your wristwatch as backup. Where in the world you are working on the review determines the pace of life going on outside your office window. In tune with stereotypes but based on research, if you are living in Dublin, Berne, Zurich or Frankfurt you will be living at a faster pace than in Rio de Janeiro, Jakarta or Mexico City, and presumably also feeling more time-pressured and frenetic.

If you are distracted while reading this, you will have time to generate an entire assortment of thoughts oriented either predominantly to the present, future or past, depending on your individual time perspective. If you are oriented towards the future, unless you know you have a terminal illness, an addiction, autism, or are on death row, then right now you could effortlessly zip forward mentally to any chosen point in your future and imagine which books will still be on your list of books to read before the age of 100. Now let's switch direction and go back in time. What most of us are unlikely to be able to do, due to infant amnesia, is recall the memory of our first encounter with a picture book, as no long-term memories will have been formed of these precious first pre-literary moments.

As you can see, even when you're doing some rather low-key activity like reading, a whole array of factors locate you according to lived time. At any particular point in Greenwich Mean Time, various cycles of bio-activity are ticking over (for example circadian rhythm, heartbeat, breath) and the socio-historical convention of a calendrical system locates you according to a particular date. You can compare your chronological age against the average life expectancy for your country of residence, locate yourself according to generational time, and know whether or not you are transmitting time cues (biological, social and cultural) to the next unborn generation. Any drug, alcohol or food intakes have a negligible immediate influence on your future longevity. Your muscles, speech and social exchanges are organised according to various temporal features including cyclicity, pace and rhythm. You are also placed by whether or not you are relating to time polychronically, and what category of time use you are currently in. Your attitudes to time influence whether your thoughts are past-, present- or future-oriented, as well as your perception of time pressure.

Singling out relationships with time as applied to reading not only serves to show how many relationships with time coexist while carrying out such an ordinary activity, but also draws attention to the levels of involvement – brain, behaviour, society and culture – which determine our relationships with time. One notable absentee from this line-up is genes, but I'll return to them in a bit.

If we stick to this roll-call approach, we end up ignoring how the quality of our relationships with time varies across our lifespans. And so let's now trace the nature of those relationships across the course of our lives. I start with cultural influences and then move on to biological influences, including those which take into account genetic variation and the visibility of the body's circadian rhythm and associated patterns of sleep–wakefulness.

Relationships with time across our lifespans

Even in utero, cultural factors are already exerting their influences on time. We know this because languages are rhythmically different and newborns can differentiate between them, showing that they are familiar with the rhythm specific to their native language culture. In addition to this early exposure to the rhythm of language, our thoughts, memory and communications are influenced by parental speech and gestural input in early childhood. For instance, earlier I showed that autobiographical memory may extend back earlier in life in Europeans and Caucasian-Americans due to their child–parent interactional style being more child-focused and emotional than interactions between, say, Chinese parents and their children. This particular combination of self-focus and emotion is a potent cocktail in terms of the effect it has on the brain's capacity to encode events.

Other insights into how cultural influences affect the way children start to relate to time come from those who have compared traditional societies with contemporary ones. The Nuer of Sudan measure time by making references to events rather than clock time. They learn this from an early age, and so the date of a flood or a war will be related to 'when my calf was so high' or 'when so and so was circumcised'. Another example of alternative conceptions of time comes from Rafael Nunez at the University of California, San Diego. Nunez showed that an Amerindian group called the Aymara have, by our standards, a reversed sense of time. In English, Japanese and other Indo-European languages, children learn to think of the past like adults do, as behind them. For instance, we gesture behind us for the past and in front of us to represent the future. In contrast, Aymaran children gesture in front of them to talk about the past, and behind their backs to refer to the future. Their word for past is *nayra*, which means eye, sight or front, whereas their word for future is *q"ipa*, which means behind or back. These stark divergences between cultures show both how arbitrary our relationships with time are, and also how influential culture actually is in how we treat it.

Symbolic of the child's rite of passage into clock culture is the acquisition of their first wristwatch. In Great Britain Early Learning Centres sell 'teach the time' wristwatches intended for children from the age of four upwards. Well before starting school, children get used to seeing adults frequently glance at their wrists. If back-to-back scheduling, punctuality and fast-paced living are the default lifestyle around which family routines, schooling, play and bedtimes revolve, then following these early foundations, these relationships will continue to have a powerful grip throughout life. This is especially the case if the various roles of adulthood – employee, consumer, holidaymaker and so on – continue to echo the template of these earlier experiences. At the other end of the continuum, children brought up predominantly on a diet of event-based living will treat time differently because they do not need to constantly check progress against a schedule. The effort of making cognitive estimations of duration is irrelevant to cultural norms, and therefore they do not call on their brains to engage in these routine calculations. This is a persuasive example of the way in which different cultures draw differentially on mental resources, making different kinds of demands on the brain. This is no more relevant than in the twenty-first century in which advances in information technology have created new relationships with time. In Japan, teenagers playing computer games have developed super-fast decision-making abilities, acting on multiple sets of information. This reveals the plasticity of our brain to handle information both in parallel and serially but also highlights the way the defining features of our temporal landscapes can vary across different historical eras.

Talking about a constant immersion in clock culture implies that it is immutable and ignores the transitions made when people seek out new life directions. Physical relocation or a career change can transform the quality of relationships with time. Take the monk who quits the prayer–meditation–chores cycle of monastic life in order to become a city stockbroker or vice versa. Or perhaps more topically, think of the tenth of the UK population who will move abroad by 2010, quite possibly not only relocating to a new home but also choosing to divorce themselves from their current pace of life.

As none of us chooses our biological parents, we are currently unable to alter the genetic bases of our relationships with time. At this point in the early twenty-first century, when the race is still on to identify genes relating to specific relationships with time – be they disease, longevity-enabling or clock genes – certainly no one is yet in a position to build designer human babies selected for desirable relationships with time. In any case, a critical point, especially in the case of longevity, is that this outcome is determined by interactions between genes as well as the suitability of conditions in utero, not to mention parental age and a host of lifestyle factors, which work collectively. In the case of biological timekeeping, we already know from animals that it is possible to alter the speed of the circadian rhythm by removing or altering genes and there is no reason why this sort of genetic manipulation should not at least be technically possible in humans too. But given that humans living in the twenty-first century – more so than in any earlier century – live among temporal diversity, this too needs consideration alongside the constraint or advantage of genetic influences. Temporal diversity includes varying levels of artificial light intensity, diverse patterns of night working, variation in the capacity to cross international travel zones and different access to pharmacological intervention. These all affect the patterns of our rest–activity. As more is understood about genes, their interactions and the impact of pre-natal conditions, we will comprehend more about the foundations of our timekeeping and how this interacts with social and cultural influences across our lifespans.

One relationship with time showing variation, particularly at the very beginning and towards the end of the human lifespan, is the circadian rhythm. After a full-term birth there is usually a delay of a few months until the circadian rhythm of the baby becomes detectable. Thereafter, if they are fortunate, parents will find a regular sleep–wake cycle emerges around the age of six months. In adolescence there is a phase delay to the circadian rhythm mediated by the dual effects of hormones and late-night socialising. The timing of melatonin secretion is associated with maturation in puberty, and the delay in secretion that comes at this time impacts on the timing of the sleep–wake cycle. Fast-forwarding to the other end of the lifespan, with age the crystalline lens of the eye transmits less light, which flattens the amplitude of the melatonin rhythm. With the number of people in the UK aged over sixty expected to rise to 20 per cent over the next fifty years, an entire population of older shift workers will be in danger of reduced-quality sleep and the health costs that this brings. Not only that, but there will be more cases of people with Alzheimer's disease, with the dual problems of memory loss and night-time disturbance due to damage to the various neuronal pathways responsible for maintaining the march of the circadian rhythm.

We see that changes to the circadian rhythm are greatest right at the beginning and towards the end of human life, and this is explainable by immature neuronal pathways, and then damage to, or disease of, pathways which were once fully functioning. There is a risk, however, that by focusing exclusively on the changes that occur to the brain, we downplay the potential influence of social factors on sleep–wake cycles in adulthood. In the twenty-first century, the increasingly 'anytime' attitude to consumption in North America, western Europe and other cities worldwide constitutes a potential time bomb of health problems for the future. It is lower-paid workers who are more likely to be working through the night, and therefore they are more likely to suffer health inequalities, including greater risk of heart attacks. As we are discussing inequalities, I'll pull together the pick of the worst examples this book has revealed, as well as providing some signposts for ways to campaign for change.

Unequal relationships with time

Our most fundamental relationship with time, our own longevity, is also the world's most unequal. Whilst longevity continues to nudge up in industrialised countries, by 2010 the size of the gap between the world's lowest and highest national average life expectancies will likely span over fifty years. The HIV/AIDS virus, together with mass poverty, where one billion people worldwide live on the equivalent of $1 a day, curtails longevity.

These massive inequalities are paralleled by comparable disparities in financial well-being. US pet owners spend around $11.6 billion annually on dog and cat food. This falls only $2 billion short of the amount that would be needed to provide basic health and nutrition for every person in the world. View this figure alongside the fact that in affluent societies where food is plentiful there is a virtual epidemic of heart disease, cancer, diabetes and other health conditions associated with obesity, with a consequent risk of reversal in previously ever-increasing lifespans. For instance, 65 per cent of Americans are overweight or obese, and the average weight of American adults has increased by twenty-five pounds since 1960, with height increase only accounting for a small proportion of the rise. Portion sizes have increased too. Famously, a 1950s McDonald's adult meal was a hamburger, small fries and a small Coke, whereas this is now a Happy Meal™, aimed at children. Yet worldwide an entire research community exists to assess the contribution of lifestyle factors in determining longevity. Given the obesity crisis, people are clearly reluctant to change their eating habits (after all, they know what is bad for them but continue to eat it), which tells us that in affluent societies people's lifestyles are heavily invested in a present time perspective.

Although financial aid is obviously absolutely critical for beating poverty and famine, this alone will not alter the poor's most fundamental relationship with time. Aside from the problem of some African countries having notoriously corrupt and incompetent leaders, exports from developing countries face tariff barriers which amount to twice as much as they receive in aid. Such countries must also cope with the economic and social power of transnational companies. A startling fact is that only twenty-five countries in the world are known to have a larger gross domestic product than the annual value of sales of General Motors, the world's biggest transnational corporation.

The African Commission of Human and Peoples' Rights has highlighted the fact that Africa is rich in natural resources like copper, gold and diamonds, and also in human resources. But take the case of Ghana, where one-third of the population still depends on untreated spring or river water, 53 per cent of heads of household have never spent any time in education and only 61 per cent have access to health services, with availability worse in rural areas. There are more Ghanaian doctors in New York than there are in the whole of Ghana. Africa sends 77,000 professionals abroad each year to work, a drain of skills which directly reduces longevity.

Another factor which influences long-term developments is the time perspective of the 191 United Nations member states. In their formation of the UN millennium development goals, they set as their target the year 2015 by which to achieve universal primary education, reduce child mortality, and combat HIV/AIDS, malaria and other diseases. In terms of this vision, one finding from the research of Nancy Petry is relevant. Obviously no progress can come too soon and the setting of short- and mid-term goals is imperative. Nevertheless, Petry's work showing that addicts project an average of eighteen years into the future demonstrates that the leaders of the countries of the world have an even more present-oriented perspective than addicts. Such a limited future time perspective clearly has negative implications for longer-term problems such as global warming. This contrasts with the approach taken in harmonic time cultures, where the focus is as far ahead as two generations, as in the case of trees being planted for unborn grandchildren. It is realistic to say that people's relationships with time as expressed in terms of longevity will only be transformed if, alongside efforts to increase aid, the governments of the world's most powerful nations are prepared to enforce strategies which redistribute wealth and resources using time perspectives substantially more future-oriented than they are now.

Another unequal relationship with time relates to the extent of excessive working. Compared to Europe, where in some countries there are directives on working hours and vacations, in some parts of the world there is still astounding poverty of time, particularly in sweatshops in China, where adults slave round the clock, often in poor light conditions. Some responsibility for this situation must lie with the consumer who chooses to buy the products of sweatshops, which is why the work of pressure groups and aggressive awareness-raising campaigns is so important.

However, even in countries that prohibit sweatshop working, the rise of the anytime culture ensures inequality for immigrant and other low-paid workers. The very concept of unsocial working hours is disappearing; night work is normal duty. Given that this expectation is company rather than employee driven (unless it happens to suit an employee to work nights), arguably there will be a doubly negative impact on health, due to the negative effects of work hours being set by the company rather than by the employee as well as all the usual negative health outcomes from doing night work on a long-term basis. We already know that night working in itself increases the likelihood of coronary heart disease by 40 per cent. The British Heart Foundation estimates the cost to the UK of heart disease due to all causes at over £7 billion, with £3 billion of lost earnings to the economy and 400 million hours of care annually for patients. It remains to be seen how the portion of these costs associated with night working compares to the financial fruits of the night economy. Given that currently one in seven people working in the UK does so between 6 p.m. and 9 a.m., whereas in 2020 this is expected to rise to one in four, if governments are seriously interested in preserving and enhancing our longevity, then they must exert pressure on employers to improve working conditions. For example, as night workers often have no access to canteens serving hot food, they tend to eat more cold snacks; employers need to provide alternative immune system bolstering options when canteens are closed. They also need to offer health checks. As for the social impact, 58 per cent of people questioned by the Future Foundation believe that night work is detrimental to family life, and we already know that the divorce rate is higher in these families.

Having discussed general inequalities, what about those people with the most grossly deprived relationships with time? In terms of basic survival, progress on under-five mortality was slower in the last decade than in any since 1960, a consequence of the long-standing problems of lack of access to healthcare including immunisation, adequate nutrition and safe water. Tragically, there has also been little change since the last decade in lessening the persistent inequalities in children's use of time. Some 120 million children spend no time in primary-level education. Lack of government investment in schools and infrastructure, low income, prejudice and discriminatory practices all contribute. The average spent on education amounted to less than 4 per cent of the gross domestic product in sub-Saharan Africa and as little as 1.6 per cent in Ghana (1981 figures). In general, the more time girls spend outside school labouring, the more this affects the body's composition in terms of the ratio of lean mass to body fat, which potentially adversely affects the survival chances of the next generation. In this way, negative outcomes fall like a line of dominoes:

there is a trail of destruction across different relationships with time, across different generations.

For the 300,000 children worldwide who are soldiers – many of them under ten – their use of time bears no resemblance to that of children growing up in leisured society. Impoverished children are seduced by the lure of food, clothes and money into a life of violence, rape and killing. Save the Children and Unicef are active in aiding the release of child fighters, and so far in Congo their collective efforts have demobilised 5,500 child soldiers. One of the huge challenges faced by such aid organisations is to transform the mindset of children who have known only hostility, death and getting what they want using guns. Children who have returned from war spend time in transit camps and opportunities are created to aid their reintegration by participation on vocational training programmes. Massive injections of support and resources are still urgently needed in these areas to rehabilitate the remaining 294,500 child soldiers.

Yet another inequality for youngsters emerges from a statistic: there are 100,000 blind children in French-speaking African countries. Horrifyingly, around half of these die within a couple of years of losing their sight. The main causes of blindness are corneal scarring due to measles, vitamin A deficiency, conjunctivitis of the newborn and harmful traditional eye medicines. Not only does this tragedy constitute a glaring inequality of time, but as I described in Chapter 7, blindness forces a person to develop relationships with time and space that are different to a sighted person's. Even in societies that are supposedly disability aware, the adaptations necessary for such new relationships are not adequately taken into account in public spaces. For instance, one challenge faced by people with a visual impairment is knowing when they reach the top of an escalator and a localised announcement would be simple to install, but this is not done. More serious are the consequences for the blind in illiterate societies with no provision for Braille education. On the positive side, the elimination of avoidable blindness by 2020 is a goal currently being tackled jointly by African governments, international bodies and non-governmental development organisations through interventions like introducing vitamin A-rich sweet potatoes into children's diets.

Women form another group experiencing inequalities with time. Although advances in contraception have enabled some women to take control of their reproductive potential, worldwide there are 350 million women who want, but who do not have access to, modern methods of contraception; the work of Marie Stopes International Worldwide is crucial in campaigning for change.

Another area of inequality for women in industrialised societies is found in the context of the work–life balance. In some senses, the term is unfortunate because it overlooks the increased fragmentation of time slots when living in a scheduled society. This problem has been exacerbated by some male sociologists who have restricted comparisons between men and women, instead lumping together time spent in different activities. The typical schedule – caring for older relatives, the school run, after-school activities, shift work, commuting, shopping and so on – can result in sleep deprivation for women worse than that suffered by the previous generation. Working mothers with babies often have little over three hours sleep, some two hours less than their parents did. Any kind of back-to-back schedule causes a significant cognitive load on the brain, and this is particularly the case with working mothers (and other carers – for example, in the UK, one in four women over fifty has care responsibilities for the elderly, sick or disabled) for whom in a typical day there are many transitions between activities handled in sequence and in parallel. Because, by definition, a back-to-back schedule contains numerous transitions between slots, it demands monitoring anticipated progress against the clock. And so, in order to keep on schedule, it is also essential to have a keen sense of time's passage. Monitoring the clock and being aware of time impose a high memory load and require attention, which prevents the mind fully concentrating on conversations in hand, which in turn presumably leads to mistakes and contributes to an overwhelming sense of time pressure.

There is plenty of scope for innovations to overcome these cognitive constraints and take the edge off time pressure for women. The introduction of portable voice recognition memory aids into the workplace would allow the constantly running home and work 'to do' lists to be edited 24/7. Setting a personal vibrating alarm to go off at a chosen interval before the end of a meeting would provide a subtle time cue to wind things up.

Different kinds of scheduling challenge occur for children whose parents live apart. The parent who lives with the kids during the school week often reports feeling that the other one has the better deal – movies, shopping trips, McDonald's and a chance to watch a football match – while they handle the laundry, school runs and making packed lunches. Of relevance here is the UK government white paper 'Supporting Families' (Home Office, 1998), which suggests that more family time means more 'quality time' for parents to spend with their children. For families who live apart, the notion that 'quality time' comes from time spent together as a family is rather exclusionary. Not only that, but Pia Christensen from the University of Copenhagen believes that underlying this notion is an assumption that family time is 'good' for children. After listening to the views of English children living in the north of England about time with their families, Christensen found that these ten- to eleven-year-olds did *not* actually want more time with their families. Rather, she identified five alternative kinds of time that did matter to them.

1. Ordinariness and routine

Children valued a shared family teatime, sitting on the sofa watching TV or other activities which brought the family together. Teatime marked a break between playing outside and family time.

2. Someone being there for you

What mattered to children was someone being available when it mattered. They wanted to know that somebody could help, do things for them or stick up for them. Children were not asking for more time with their parents; they just wanted to know that parents were there for them.

3. Value of having a say over time

Children viewed family time as the time they spent together as well as the time they were able to spend under the same roof but spent alone. They valued making decisions about how to use their time – when to choose to share space with younger brothers and sisters and when to have privacy.

4. Having peace and quiet

Children saw their everyday time as under pressure from external factors over which they had minimal control. This was difficult when they were looking out for younger brothers and sisters or when they were being minded at another person's house under a different set of expectations about time use.

5. Being able to plan time

Having homework signified a new and regular demand on children's time which could be difficult to organise, with younger brothers and sisters pestering them, after-school jobs, chores, friends and activities. Children living in rural areas had to bid for time to see their friends, often depending on their mothers being available to transport them, whereas those living in urban areas were able to decide how to use their time more independently.

Increased participation in the labour force has lead to varying consequences for women worldwide. One solution to the problem of juggling the dual roles of caring for elderly parents and motherhood adopted by Japanese women has been to remain single and childless. Contributing to this trend is resentment at the virtual absence of hands-on parenting by Japanese fathers. Japanese husbands with at least one child under six spend an average of only twenty-one minutes a day with their offspring and the societal expectation is still that women should do the bulk of the housework, childcare and part-time work. With a view to slowing the plummeting birth rate, the Japanese government has pressured businesses to limit overtime in order to give men more time to spend with their children and get busy making larger families. Some administrative authorities give one million yen (£5,000) to a couple who have a third child, and discounts at shops and restaurants.

Efforts to ease the conflict between work and caring responsibilities have been firmly on the agenda of the Work Foundation in the UK. One example of a success story from the point of view of both employer and employee comes from British Telecom (BT). BT has created eBT, allowing employees to complete work in multiple locations. Around 70 per cent of training is now delivered online. The business benefits have included improved staff retention, reduced absenteeism, increased productivity, happier customers and reduced costs. Encouragingly, there are also resources available for managers to create a flexible work environment, which reduces the initial investment of time needed to transform the temporal practices of the workplace.

Despite the ways in which mobile technology has improved our lives, one downside is the risk that certain kinds of decision-making and problem-solving have become predominantly impulsive rather than benefiting from a period of reflection, incubation or even 'timelessness'. In a state of timelessness, a person becomes settled in the enjoyment of a present-moment activity, engaged in the here and now, and capable of deep concentration without being conscious of time's passage, the future or the past. The idea is to stay wrapped up in the unfolding present moment, immersed in 'flow'. A study into creativity in the research and development division of a US scientific corporation showed that the more frequently workers experience timelessness, the more creative they become. This highlights the importance of designing workplaces so that they minimise stress, alienation and boredom and instead create opportunities for workers to develop solutions and skills as well as receive job satisfaction.

One last unequal relationship I want to mention is people's diversity in relating to time. As we have seen, in an automated society everyday features of the built environment like lift doors do not cater to people's differing pace of life. As you move about in your home, office, transport, park or shopping centre today, you may spot mismatches or good fits between your own body and your temporal landscapes, and notice how these could have been customised.

Given that in European countries three-quarters of employees work outside the traditional nine to five, it is surprising that the majority of diaries sold in high-street chain stores typically only still run from 7 a.m. to 7 p.m., with space allocated to the rest of the night being rare. Alongside the urgency of monitoring the health of night workers, the colonisation of the night provides new opportunities for commercial ventures, including healthcare initiatives such as the introduction of new food products which could help to optimise the nutritional intake of night workers. Helplines could also be set up to give advice and support to night workers in order to improve their health and lifestyle.

By taking into account the variety of ways in which we act in time, we are better able to design public spaces and the built environment to our advantage, and this may even inspire inventions too. In turn, as we have seen throughout the book, innovation creates new possibilities for altering the quality of our relationships with time.

How innovations have transformed our relationships with time

It was only 150 years ago that wristwatches first came into use, following the earlier innovation of factories and the growth of the railway networks. These developments collectively popularised punctuality, which explains the degree of scheduling associated with clock culture. The invention of the wristwatch ensured that people living in clock time cultures developed highly practised cognitive skills in timing short intervals, while relying on the brain's internal timekeeper to ensure that they stuck to the demands of their schedules. These skills also help in other everyday situations where there is no need to check an interval against your watch, but nevertheless skills of estimation of duration are called upon. For instance, when a traffic light hits amber, a quick decision is needed as to whether or not you can squeeze through before the red. As there is already a mental representation of how long the amber phase lasts, this acts as the comparison point and informs whether you decide to brake or not.

Although we fare well at estimating short intervals, as I showed in Chapter 1 we are notoriously bad at judging how many hours spare time we have had in a particular week, as our attention is devoted to ongoing activities rather than tracking the duration of specific kinds of creativity. A useful invention to overcome the constraints of the mind's limited attention capacities would be a time log built into your mobile which allowed you to audit your time use over any specified period. As well as being programmed with standard activities like showering, driving and eating, it would have free slots enabling you to customise it with your own activities. It could deal with multitasking by switching between simultaneous activities at the touch of a button. Instead of simply having a subjective notion of how your time was spent you would have an objective record which would allow you to work out how to make changes. It would also help you to respond accurately to requests to account for time spent on different kinds of work activities.

Another notable historical breakthrough occurred around 120 years ago when strong artificial light became widely available domestically, allowing people the choice of which portion of the light–dark cycle to colonise. Of course, it is only relatively recently that a genuine round-the-clock lifestyle has become feasible and then only in urban areas, and I have already explained how colonisation of the night offers freedom of choice for the rich, but adversely affects the health of those at the other end of the scale.

About 100 years ago the first powered flight set the stage for passengers of the future to travel across different time zones. Now growth in the aviation industry is set to rise by 5–7 per cent per year. Although this will bring the opportunity for even more people to cross international time zones for business or pleasure, it also comes at a cost to health and well-being in a number of ways. We know that jetlag works against the circadian rhythms, and that it actually takes the liver up to two weeks to adapt. This shows how our bodies and brain are only programmed to make gradual adjustments and cannot keep pace with rapid travel. The effects of jetlag are known to be ameliorated by a dose of melatonin and exposure to light at strategic points in the day. A second concern arose after the introduction of jet planes in the 1970s. Because jets fly higher in the atmosphere and for longer, flight crew can be exposed to cosmic radiation for fourteen hours or more at a time. Disturbingly, incidences of malignant melanoma have been found to be two to three times higher among aircrew than the rest of the population. But these findings should be interpreted with caution – what is still unknown is whether the irregular work hours combined with the disturbances to sleep–wakefulness patterns are responsible for predisposing aircrew to health risks such as these.

Another problem with the growth of the aviation industry is that people living under flight paths will have even worse sleep than they already do. Although curfews exist on flying between 11.30 p.m. and 4.30 a.m., the reality is that more than two-thirds of British households experience night-time noise levels higher than the forty-five decibels the World Health Organisation stipulates as acceptable between 11 p.m. and 7 a.m. This level describes the sound when no one is talking and there is just a little background noise outside. Even if there are curfews, with the rise in the number of workers doing night work, those sleeping in the day will have their already shortened, poorer-quality sleep worsened by daytime aircraft noise. During the day, seventeen miles from Heathrow, planes produce seventy decibels, and seven jets thunder overhead every ten minutes. The work of local campaign groups like the Heathrow Association for the Control of Aircraft Noise is critical for ensuring change. One of their victories was the initiation of a system ensuring alteration between the runways used by planes.

There are now also other ways to bring people together. Corporate and individual business globalisation, facilitated through the internet, has led to the emergence of global virtual teams (GVTs). As a result, the boundaries between M-time and P-time cultures are becoming increasingly blurred. Take two teams based in the USA and southern Europe separated by a six-hour time difference. Their virtual team meeting could be via teleconferencing in the European team's afternoon, allowing the USA group to work on the action points in their own afternoon in time for the European group to review in the morning. The overnight gain secured by having two teams displaced in time drastically reduces the time taken to get the product to market – that is, if all goes to plan. The melange of cultures, each relating to time in different ways, could potentially sabotage prompt project delivery. Carol Saunders and her colleagues from the University of Central Florida have lined up some creative solutions for GVTs. They suggest looking at people's time styles and matching them with compatible tasks. Those closely attuned to clock time may be offered production-oriented or scheduling tasks, while people more used to living in event time are much more likely to have experience achieving goals in parallel. People living in the moment have the capacity to become so engrossed in the here and now that they are likely to come up with creative solutions to the problem in question. Of course, the danger is that using these techniques will stereotype countries or people on the basis of their relationship with time. One solution is to dedicate virtual spaces specifically for negotiating issues of individual time style. In this way, issues around scheduling, punctuality and deadlines are resolved before launching a new project.

One of the most recent transformations occurred around forty years ago when the home freezer appeared. In the twenty-first century, the freezer allows us to create different kinds of culinary time zones, which we can switch between depending on how time-pressured we are. When time is short, easy-preparation or ready meals are any-time meals. Both the freezer and the microwave have contributed to an erosion of time boundaries for joint family activities. Although some view the demise of the family meal as a sign of breakdown in societal values, collective eating could now simply be low on the family wish list and parents may prefer to do other activities together. An alternative culinary time sector is the relaxation zone – the opportunity to have a leisurely weekend meal with friends or even to live according to the ethos of the slow movement I mentioned in Chapter 1.

The wristwatch, domestic lighting, the passenger plane and the freezer have all altered the choices available for us in the ways we relate to time. In future equivalent time frames, further technological advances may again present new options. As it is, newspapers are already printing project schedules for anticipated tourist trips to Mars – which will require a new public calendar to take into account the extra thirty-seven minutes in each Martian day. And in relation to another of our relationships with time – the organisation of our lifespans – given recent medical advances in ovarian transplants, together with rising longevity, there could eventually be new options for when reproduction takes place, which would allow more women the scope to combine a career with having children while caring for elderly parents.

Moving beyond the nine to five

In this, the last word about your life in time, I invite you to catapult yourself beyond the nine to five. Why? Well for a start, the concept is now outdated; in the twenty-first century we live in a flexi-time, flexi-place economy. Secondly, having a mindset revolving around the nine to five prevents you moving other relationships with time up the agenda, those which potentially offer a better quality (or quantity) of life. Not all of us have access to the same freedom of choice – as I have already argued, genes, age, location, lack of funds, illness, inequality or other circumstances can all disadvantage people. Nevertheless, hopefully you will agree that there is still scope to make you time-wiser or time-richer (or both), depending on what values are important to you. The questions that follow are intended to prompt some personal evaluation of your own relationships with time.

- Are you sufficiently temporally mobile – do you have the scope to choose which portion of the light–dark cycle to work in, whether to relocate to a new pace of life or to switch to a workplace with more employee-driven flexibility?
- How susceptible are you to advertisements telling us how we *should* or *could* relate to time?
- What slogan best represents your ideal type of relationship with lived time?
- Are you sufficiently well off to be able to afford a diet rich in antioxidants, and to be able to purchase all the time-saving gadgets that living in a leisured consumer society brings?
- To what extent do time-saving gadgets allow you to multitask compared to what was possible in the Victorian era and how many of the following do you normally

run at the same time: iRobot's Roomba vacuum cleaner, washing machine, dishwasher, stereo, phone, bread-making machine, fan and freezer?

- Does the particular culinary time zone you inhabit maintain the same pace, regardless of which day of the week it is?
- How concerned are you about the accuracy of timepieces in your house? Do you have a clock which picks up the official British Standard frequency time, overseen by the National Physical Laboratories atomic clock and broadcast by the British Telecom International Transmitter at Rugby?
- Are there ever situations where you question the importance placed on punctuality?
- How well do you keep track of short time intervals?

When it comes to longevity does anyone in particular inspire you? Perhaps supercentenarians like Jean Calment, the Okinawans or the Seventh day Adventists? Alternatively, perhaps someone you know is terminally ill, and you are starting to become aware of their altered time perspective and how they value making each day count? Given that you live in a society where people are expected to live decades beyond retirement, is there an age which you routinely mentally fast forward to? How many different generations of your family are alive and what is the spacing between generations? Is your body's current level of exposure to pollutants going to manifest itself in four generations time?

How does your vacation allowance compare with that taken globally? Can you afford to fly to different time zones and overcome the effects of jetlag as quickly as possible? Next time will you take a dose of melatonin to ease the transition to local time? Do you wear a watch and live according to clock time while on holiday? While on holiday, would you be interested in improving your game – tennis, cricket, football or whatever – by mental rehearsal of the timing of your shots? Is it possible to recreate the timelessness you experienced on vacation back in your scheduled workplace? In what ways do the demands of your workload make it beneficial to take a break from responding to the immediate demands around you – email, texts and voicemail – to nurture alternative ways of relating to time?

In your various roles as consumer, manager, resident, parent, grandparent, carer, volunteer are you concerned about influencing the choices other people have in their relationships with time? If you want to consider this, then you may find the internet resource list at the end of the book a useful starting point. And if you want to see lasting positive changes to both other people's and your own relationships with time, holding a future time perspective will serve you well.

NOTES

Trademarks were correct at time of going to print. This section lists research sources, cued by page number and identifying phrase.

Introduction

Page 1. **seven minutes is about the average amount of time adults spend**—J. P. Robinson and G. Godbey, *Time for Life: The Surprising Ways Americans Use Their Time* (University Park: Pennsylvania State University Press, 1997), 323.

Page 2. **we use the word 'time' fifteen times more frequently than**—S. Johansson and K. Hofland, *Frequency Analysis of English Vocabulary and Grammar* (Oxford: Clarendon Press, 1989), 322 and 347.

Page 2. **Standard Occupational Classification**—Office for National Statistics, *Standard Occupational Classification,* vol. 2 (London: The Stationery Office, 2000).

Page 2. **the name of the world's officially recognised longest living person**—http:// guinnessworldrecords.com/

Page 3. **world life expectancy of sixty-five years**—United Nations, *World Population Prospects: The 2004 Revision Highlights* (New York: United Nations, 2005).

Page 4. **most of us appear unable to access our earliest memories**—Q. Wang, Infantile amnesia reconsidered: A cross-cultural analysis, *Memory* 11.1 (2003): 65–80.

Chapter 1

Pages 7 & 8. **Sanatogen vitamins** and **Volvic natural mineral water**—citations taken directly from packaging.

Page 8. **manufacturers of the bacteria-friendly drink Yakult remind us, 'rushing, travelling, working, stressing. how you live can upset your life'**—from advert in *The Independent,* 10 April 2005.

Page 8. **one from Orange™ tells us to 'work faster in more places',**—Orange™ Guide to the new 3G Mobile Office.

Page 8. **Vodafone Ltd. says, 'call someone; now is good'**—http://www.web factory.ie/ htm/clients/vodafone.htm

Page 8.   Microsoft Office® appeals ! 'save time: with smarter working practices to boost productivity.'—Microsoft Office® Ltd, *Smarter Working: Improve Your Team and Personal Productivity.*

Page 8.   iRobot's wire-free Roomba red vacuum cleaner—T. Pearce, 'Aye, Robot', *The Globe and Mail,* 23 October 2004.

Page 8.   the water resistant clock radio, electric tabletop wine cooler, corkscrew and automatic card shuffler—all advertised in *Innovations* catalogues (Shop Direct Group Ltd).

Page 9.   home freezer—E. Shove and D. Southerton, Defrosting the freezer: From novelty to convenience: A narrative of normalization, *Journal of Material Culture* 5.3 (2000): 301–319.

Page 9.   introduction of the 'frost-free' freezer—http://www.bosch appliances.co.uk/

Page 9.   French commentator Alexis de Tocqueville—cited by J. P. Robinson and G. Godbey, The great American slowdown, *American Demographics* 18 (1996): 42–47.

Page 9.   early ads for Coca-Cola. T. H. Eriksen, *Tyranny of the Moment: Fast and Slow Time in the Information Age* (London: Pluto Press, 2001), 56.

Page 9.   When Albert Einstein departed—Science and the pace of life, *Current Science* 1.9 (1933): 261–64.

Page 9.   The 1960s saw a new crop of books—S. De Grazia, *Of Time, Work and Leisure* (New York: Twentieth Century Fund, 1962).

Page 9.   In James Gleick's *Faster*—J. Gleick, *Faster: The Acceleration of Just About Everything* (Boston and London: Little, Brown, 1999).

Page 9.   An exception to the American hold—T. H. Eriksen, Op. cit.

Page 10.   John Robinson from the University of Maryland and Geoffrey Godbey from Pennsylvania State University gathered—J. P. Robinson and G. Godbey, Op. cit.

Page 11.   leaders of the French revolution replaced the seven-day cycle—E. Zerubavel, The French Republican Calendar: A case study in the sociology of time, *American Sociological Review* 42 (1977): 868–77.

Page 12.   Robinson and Godbey found a continual increase in the number of U.S. citizens—J. P. Robinson and G. Godbey, *Time for Life,* Op. cit.

Page 13.   in 1784 the British mail-coach service—E. Zerubavel, E., The standardization of time: A sociohistorical perspective, *The American Journal of Sociology* 88.1 (1982): 1–23.

Page 14.   Around 1850, wristwatches came on the market—R. Levine, *A Geography of Time: The Temporal Misadventures of a Social Psychologist* (New York: Basic Books, 1997), 52.

Page 14.   Characteristic of this approach to time is what Edward Hall described—E. T. Hall, *The Silent Language* (Garden City, N.Y.: Doubleday, 1959).

Page 15.   Henry Davidoff's in-depth analysis—G. Morello, Sicilian time, *Time and Society* 6.1 (1997): 55–69.

Page 16. **In his book *Geography of Time*—**R. Levine, Op. cit.

Page 16. **'Timeless' cultures are common where Buddhism or Hinduism are practised—**C. Saunders, C. Van Slyke, and D. R. Vogel, My time or yours? Managing time visions in global virtual teams, *Academy of Management Executive* 18.1 (2004): 19–31.

Page 17. **In China and other places practising Taoism and Confucianism—**Ibid.

Page 17. **Stephen Barley at Stanford University—**S. R. Barley, in F. A. Dubinskas, ed., *Making Time: Ethnographies of High-Technology Organisations* (Philadelphia: Temple University Press), 123–69.

Page 18. **Gregory Rose and co-workers—**G. M. Rose, R. Evaristo, and D. Straub, Culture and consumer responses to Web download time: A four-continent study of mono and polychronism, *IEEE Transactions on Engineering Management* 50.1 (2003): 31–44.

Page 19. **According to Robert Levine—**R. V. Levine and A. Norenzayan, The pace of life in 31 countries, *Journal of Cross-Cultural Psychology* 30.2 (1999): 178–205.

Page 19. **Levine compared the pace of life—**Ibid.

Page 21. **There are signs that collectivism starts young—**J. A. M. Farver, B. Welles-Nystrom, D. L. Frosch, S. Wimbarti, and S. Hoppe-Graff, Toy stories: Aggression in children's narratives in the United States, Sweden, Germany, and Indonesia, *Journal of Cross-Cultural Psychology* 28 (1997): 393–420.

Page 22. **Famous psychologist Stanley Milgram—**S. Milgram, The experience of living in cities, *Science* 167 (1970): 1461–1468.

Page 22. **the Centre for Future Studies forecast—**http://www.network54.com/Forum/thread ?forumid = 257194&messageid; = 1093820948

Page 22. **now stands at over 80,000 members in over 100 countries—**http://www.slowfood.com/eng/sf

Page 22. **small group of leftists in 1986—**Ibid.

Page 23. **in protest against the plans for a McDonald's—**M. Schneider, Tempo diet: A consideration of food and the quality of life, *Time and Society* 6.1 (1997): 85–98.

Page 23. **One of the latest projects is the Ark of Taste—**http://www.slowfood. com/eng/sf

Page 23. **Some 605 hours—**UBS Prices and Earnings Survey (2000) from *Realtime*—one-off supplement (not dated) in association with *The Observer*, 2004.

Page 23. **Aubry laws, which saw the introduction of the 35-hour week—**J. Pélisse, From negotiation to implementation: A study of the reduction of working time in France (1998–2000), *Time and Society* 13.2–3 (2004): 221–44.

Page 24. **Table 1 adapted from—**UBS Prices and Earnings Survey (2000). Op. cit.

Page 24. **recruitment consultancy Adecco—**Microsoft Office® Ltd, *Smarter Working: Improve Your Team and Personal Productivity*, 5.

Page 24. **for as little as—**N. Klein, *No Logo* (New York and London: HarperCollins, 2001).

Page 24. **The bold opening statement of the European Commission's**—European Commission, Time use at different stages of life: Results from 13 European countries. Theme 3: Population and social conditions, *Working Papers and Studies* (July 2003).

Pages 24–25. **employed European woman spends 25 more minutes per day juggling**—Ibid.

Page 25. **Table 2 shows the situation**—G. Richards, Vacations and the quality of life: Patterns and structures, *Journal of Business Research* 44 (1999): 193, with permission from Elsevier.

Page 25. **In the Netherlands some employers run schemes**—Ibid., 195.

Page 26. **many only take as few as 8 days**—Ibid., 194.

Page 26. **there were 63 official cases of *karoshi***—Ibid., 195.

Page 26. **the Japanese government has taken steps**—R. V. Levine and A. Norenzayan, Op. cit.

Page 26. **a fifth of the U.S. adult population choosing to adapt**—S. Brittain, 'There's More to Life Than Growth', *Financial Times* (Japan), March 26, 2004.

Page 26. **Figure 1 shows that**—Figure based on data extracted from three sources: (1) J. Gershuny, *Changing Times: Work and Leisure in Postindustrial Society* (New York: Oxford University Press, 2000), 163; (2) J. P. Robinson and G. Godbey, Op. cit., 95 and 126; and (3) European Commission, Time use at different stages of life: Results from 13 European countries, Theme 3: Population and social conditions, *Working Papers and Studies* (July 2003).

Page 27. **Some 85 per cent of U.S. workers**—R. V. Levine and A. Norenzayan, Op. cit.

Page 27. **Norwegian government reports that sick**—T. H. Eriksen, Op. cit., 126.

Page 27. **That the French are found to sleep**—European Commission, Time use at different stages of life: Results from 13 European countries. Op. cit.

Page 28. **Dale Southerton from the University of Manchester talks about**—D. Southerton, 'Squeezing time': Allocating practices, coordinating networks and scheduling society, *Time and Society* 12.1 (2003): 5–25.

Page 29. **Between a third and a half of British school children have part-time work**—M. Leonard, Working on your doorstep: Child newspaper deliverers in Belfast, *Childhood* 9.2 (2002): 190–204.

Page 29. **young newspaper vendors in Ethiopia**—M. Woodhead, *Children's Perspectives on Their Working Lives* (Stockholm: Rädda Barnen, 1998).

Page 29. **A project by Martin Woodhead**—Ibid.

Page 30. **the work of Samantha Punch**—S. Punch, in S. L. Holloway and G. Valentine, *Children's Geographies: Playing, Living, Learning* (London: Routledge, 2000).

Page 30. **20 per cent of the world's children never have any experience of schooling**—E. Grigorenko and P. O'Keefe, in R. Sternberg and E. Grigorenko (eds), *Culture and Competence: Contexts of Life Success* (Washington, D.C.: American Psychological Association, 2004), 25.

Page 30. The remaining 80 per cent—Ibid., 24.

Page 30. Some 300,000 children in the world aged 5 to 14—A. H. Høiskar, Under age and under fire: An enquiry into the use of child soldiers, 1994–1998, *Childhood* 8.3 (2001): 340.

Page 31. An inquiry by Astri Halsan Høiskar—Ibid., 340–60.

Page 31. Probably the world's largest concentration of child labourers—K. Bales, *Disposable People: New Slavery in the Global Economy* (Berkeley and London: University of California Press, 2000), 200.

Page 31. Some children spend between eight and ten hours a day—E. Grigorenko and P. O'Keefe, Op. cit., 37.

Page 32. Glover Ferguson, chief scientist at Accenture Ltd—*The Globe and Mail*, 15 October 2004.

Page 32. 'The mass timetable of the industrial world, of the "9 to 5" office world . . .—I. Hardill and A. Green, Remote working—Altering the spatial contours of work and home in the new economy, *New Technology, Work and Employment* 18.3 (2003): 212.

Page 32. The latest European survey (based on 15 countries) showed that only 24 per cent of people now work office hours Monday to Friday—G. Costa et al. Flexible working hours, health, and well-being in Europe: Some considerations from a SALTSA Project, *Chronobiology International* 21.6 (2004).

Page 32. The changes in shopping hours in northern Europe—Harvey et al. *24-Hour Society and Passenger Travel: Final Report* (Time Use Research Program: 1997).

Page 32. In Britain Sunday trading became legal a decade ago—S. Ryle, 'Ten Years That Shook the Tills', *The Observer*, 29 August 2004.

Page 33. In Sweden the X2000 high-speed train and Norwegian equivalent, the Signatur—http://europrail.net/canada/info/high-speed.html

Page 34. this has been shown to lead to greater incidence of—G. Costa et al., Op. cit.

Page 34. the average British couple spend—M. Frith, 'Couples Starved of Social Time, Study Finds', *The Independent*, 16 July 2004, 9.

Page 34. In Britain around 1 in 10 women—E. Agree, B. Bissett, and M. S. Rendall, Simultaneous care for parents and care for children among mid-life British women and men, *Population Trends* 112 (National Statistics 2003): 29–35.

Page 35. In Europe the highest proportion of shift workers—T. Kauppinen, The 24-hour society and industrial relations' strategies, *European Industrial Relations Association* (June 2001).

Chapter 2

Page 37. electric lighting was first installed for domestic use in the 1880s—L. Kreitzman, *The 24 Hour Society* (London: Profile Books, 1999), 23.

Page 37. As some 20 per cent of employees now work by artificial light—T. Kauppinen, Op. cit.

Page 38. Now statistics from around Europe suggest that since 1995 the number of people working shifts has increased by 7 per cent—Ibid.

Page 38. Nova Scotia continues to ban—Canadian Press, 'Nova Scotians Refuse Sunday Shopping', *The Globe and Mail,* 18 October 2004.

Page 39. call traffic on British Telecom at 4.30 a.m. has increased by 400 per cent— L. Kreitzman, Op. cit., 10.

Page 39. looked into the distribution of sleep patterns across six countries—Harvey et al., Op. cit.

Page 39. Flies—S. M. W. Rajaratnam and J. Arendt, Health in a 24–hour society, *The Lancet* 358 (2001).

Page 39. small mammals—P. D. Penev et al., Chronic circadian desynchronization decreases the survival of animals with cardiomyopathic heart disease, *American Journal of Physiology* 275.6 (1998): H2334–H2337.

Page 39. claimed to be equivalent to smoking 20 cigarettes—L. Kreitzman, Op. cit., 116, cites D. C. Whitehead, H. Thomas, and D. R. Slapper, A rationale approach to shift-work in emergency medicine, *Medicine* 21 (1992): 1250–1258.

Page 39. night shifts on a regular basis increases the risk of heart disease in humans by as much as 40 per cent—H. Boggild and A. Knutsson, Shiftwork, risk factors and cardiovascular disease (reviews), *Scand. J. Work Environ. Health* 25.2 (1999): 85–99.

Pages 39–40. workers tend to graze on a series of cold snacks—J. Waterhouse, Measurement of, and some reasons for, differences in eating habits between night and day workers, *Chronobiology International* 20.6 (2003): 1075–1092.

Page 40. the risk of peptic ulcers is between 2 and 5 times higher for night workers— K. Segawa, S. Nakazava, Y. Tsukamoto, Y. Kurita, H. Goto, A. Fukui et al., Peptic ulcer is prevalent among shiftworkers, *Dig. Dis. Sci.* 32: 449–53.

Page 40. The conclusion that women's reproductive health—P. H. Gander, Sleep, health, and safety: Challenges in a 24–hour society, *Thomas Cawthron Memorial Lecture* No. 55 (New Zealand: Cawthron Institut, 1997).

Page 40. The sleep debt resulting from the cumulative effect of night working causes a catalogue of—K. K. Papp et al., The effects of sleep loss and fatigue on resident-physicians: A multi-institutional, mixed-method study, *Academic Medicine* 79.5: 394– 406.

Page 40. Morning shifts that start before 6 a.m. have been shown to lead to a higher risk of severe sleepiness than—M. Ingre et al., Variation in sleepiness during early morning shifts: A mixed model approach to an experimental field study of train drivers, *Chronobiological International* 21.6 (2004): 973–90.

Page 41. morning types are also more likely to have a lifestyle—T. Monk et al., Morningness–eveningness and lifestyle regularity, *Chronobiology International* 21.3 (2004): 435–43.

Page 41. **The discovery that body temperature**—R. G. Foster and L. Kreitzman, *Rhythms of Life: The Biological Clocks That Control the Daily Lives of Every Living Thing* (London: Profile Books, 2004).

Page 42. **Jurgen Aschoff from the Max Planck**—Ibid.

Page 42. **In later work by Aschoff**—Ibid.

Page 42. **According to Sally Ferguson**—http://abc.net.au/science/news/stories/ s372491 .htm

Page 43. **the risk of an accident occurring**—R. G. Foster and L. Kreitzman, Op. cit.

Page 44. **As we age, the crystalline lens**—W. N. Charman, Age, lens transmittance, and the possible effects of light on melatonin suppression, *Ophthal. Physiol.* 23 (2003): 181–87.

Page 44. **compare the 3 per cent of people**—R. G. Foster and L. Kreitzman, Op. cit.

Page 45. **If the reduction of light at**—A. Magnusson and D. Boivin, Seasonal affective disorder, *Chronobiology International* 20 (2003): 189–203.

Page 45. **Steve Kay and**—http://www.scripps.edu/news/press/081804.html

Page 47. **a century later the aviation industry**—J. P. Clarke, The role of advanced air traffic management in reducing the impact of aircraft noise and enabling aviation growth, *Journal of Air Transport Management* 9 (2003).

Page 48. **intense periods of the Iraq invasion, Modafinil was used by British and American troops**—R. G. Foster and L. Kreitzman, Op. cit.

Page 48. **The UK's Future Foundation**—J. Doward, 'UK Slogs Around the Clock', *The Observer*, 12 September 2004.

Page 48. **The Future Society projects that by 2020**—http://scotsman.com/ index.cfm?id = 1133572004

Chapter 3

Page 51. **Jeanne Calment, who died**—http://www.guinessworldrecords.com/

Page 51. **Households with three or more generations now make up 4 per cent of the USA's 105.5 million**—http://www.tucsoncitizen.com/local/ census

Page 51. **represents a growth of 60 per cent**—http://www.pbs.org/americanfamily/ gap/multi.html

Page 51. **Linda Burton**—L. Burton and P. Martin, Thematiken der Mehrgenerationen-famille: Ein Beispiel einer Sechs-bzw. Siebengenerationenfamille, *Zeitschrift für Gerontologie* 20 (1987): 275–82.

Page 53. **world's youngest mother**—http://youngest_mother.tripod.com/

Page 53. **register of the world's oldest people**—http://www.grg.org/Adams/E˙files/sheetool .htm

Page 54. **Austad has made a bet**—http://discover.com/issues/nov-03/cover/?page = 3

Page 55.   **In Rome 200 BC**—V. Marigliano, L. Tafaro, and I. Trani, The meaning of longevity in Centenarians, *The Geneva Papers on Risk and Insurance* 28.2 (2003): 238–53.

Page 55.   **In eighteenth-century France**—Ibid.

Page 55.   **Two centuries later, in developed countries**—http://www.originalghr15 .com/secrets.html

Page 55.   **At the start of the twenty-first century, the world average life expectancy**—United Nations, *World Population Prospects: The 2004 Revision Highlights* (New York: United Nations, 2005).

Page 55.   **Projected figures from**—http://www.usaid.gov/press/releases/2002/pr020708. html

Page 55.   **Take Botswana**—http://www.usaid.gov/press/releases/2002/pr020708.html

Page 55.   **increasing longevity in these areas, costs \$10**—D. Taylor, *Children and HIV/AIDS*, supplement (undated) with *The Observer*, 2005.

Page 56.   **The secretary general of the United Nations**—S. Lewis, 'UN Envoy Warns Act on Apocalypse', *Open Eye Magazine*, Spring 2003.

Page 56.   **The annual budget for the USA National Institute of Ageing is nearly \$50 million**—http://www.bsra.org.uk/HoL˙response¨BSRA.doc

Page 56.   **In Japan, the cost of setting up the National Institute for Longevity Sciences**—K. Kitani, in D. Harman, Increasing the healthy life span: Conventional measures and slowing the innate aging process, *Annals of New York Academy of Sciences* 959 (New York: New York Academy of Sciences, 2002): 517–26.

Page 56.   **Japan tops the league table**—United Nations, Op. cit.

Page 56.   **Not far behind Japan are various countries**—Ibid.

Page 56.   **Projected life expectancy in Japan in 2050**—http://www.sciencemag.org/ cgi/co! ntent/full/280/5362/395

Page 56.   **Japan also tops the league**—Y. Yamori et al., Implications from and for food cultures for cardiovascular diseases: Japanese food, particularly Okinawan diets, *Asia Pacific Journal of Clinical Nutrition* 10.2: 114–45.

Page 56.   **Between the nineteenth and twentieth centuries**—P. H. Whincup et al., Age of menarche in contemporary British teenagers: Survey of girls born between 1982 and 1986, *British Medical Journal* 322 (2001): 1095–1096. k

Page 57.   **In the case of British teenagers**—Ibid.

Page 57.   **The average age of the onset of menarche in Europe**—F. Thomas et al., International variability of ages at menarche and menopause: Patterns and main determinants, *Human Biology* 73.2: 271–90.

Page 57.   **20 per cent of the world's children had never been to school**—E. Grigorenko and P. O'Keefe, Op. cit.

Page 57.   **and that 80 per cent of the rest of the world's population only**—Ibid., 24.

Page 57. **Higher illiteracy rates are linked with**—F. Thomas et al., Op. cit.

Page 57. **Cameroon, Yemen, Somalia, Nigeria, Tanzania, Haiti, Bangladesh, Papua New Guinea, and Senegal**—Ibid.

Page 57. **In Nigeria and Ghana the average woman**—Ibid.

Page 57. **Of the 15 million babies born**—T. M. McDevitt et al., *Trends in Adolescent Fertility and Contraceptive Use in the Developing World* (USA: Department of Commerce, 1996).

Page 57. **Currently around six out of 10 women**—'Dying to Have a Baby', supplement (undated) with *The Observer*, 6.

Page 58. **In Ghana, 50 per cent of women**—T. M. McDevitt et al., Op. cit.

Page 58. **In Somalia, the average number of children is**—http:// www.bartleby.com/ 151/a30.html

Page 58. **Take Tanzania, where some 18 years ago, the number of children per family was 4.8 for urban and 7.1 for rural areas. Now it is down to 4.1 and 6.3 respectively**—Chuwa et al. (1991) and Bureau of Statistics and Macro International (1997), cited by A. Hinde and A. Mturi, Recent trends in Tanzanian fertility, *Population Studies* 54 (2000): 177–91.

Page 58. **available supply of condoms**—S. George, 'Dying to Have a Baby', supplement (undated) with *The Observer*.

Page 58. **Forty years ago the contraceptive pill**—G. Greer, *The Whole Woman* (New York: Knopf, 1999).

Page 58. **In the USA, the number of women of child-bearing age**—G. Armas, 'Record Number of U.S. Women Children', *Salt Lake Tribune*, 25 October 2003.

Page 58. **The average age of first motherhood in Britain**—http://www.stations.gov.uk/ Statbase/ssdataset.asp?vlnk = 6372&Pos = &ColRank – 2&Rank = 272

Page 58. **35-year-old degree-educated**—A. Frean, 'How Half of Women Who Expect to Become Mothers in Their 30s Are Beaten by Clock', *The Times*, 1 October 2004, 23.

Page 59. **In the UK, the rate is down to 1.6**—R. Girling, 'The Great Baby Shortage', *The Sunday Times Magazine*, 2004, 17–33.

Page 59. **In Japan, the fertility rate is also the lowest it has ever been, down to 1.28 children**—D. McNeil, 'How Japan Succumbed to a Massive Attack of Puppy Love', *The Independent*, 11 June 2004.

Page 59. **Japan Pet Food Association**—Ibid.

Page 59. **In Italy and Hungary, it is right down to 1.2**—R. Girling, Op. cit.

Page 59. **Widowed and not in contact**—W. Hutton, 'An Italian Lesson For Europe', *The Observer*, 26 September 2004.

Page 59. **robots**—D. McNeil, Op. cit.

Page 59. **20 other multi-generational black families**—L. M. Burton, Teenage childbearing as an alternative life-course strategy in multigeneration black families, *Human Nature* 1.2 (1990): 123–43.

Page 60. **Thomas Perls from Harvard**—Middle-aged mothers live longer, *Nature* 389.133 (1997).

Page 61. **Zeng Yi and**—Z. Yi and J. W. Vaupel, Association of late childbearing with health longevity among the oldest-old in China, *Population Studies* 58.1 (2004): 37–53.

Page 61. **In the annual wet season**—T. J. Cole, Seasonal effects on physical growth and development, in S. J. Ulijaszek and S. S. Strickland, eds., *Seasonality and Human Ecology*, 35th Symposium Volume of the Society for the Study of Human Biology (Cambridge: Cambridge University Press, 1993), 89–106.

Page 61. **What about negative input from**—R. Chapin et al., Off to a good start: The influence of pre- and periconceptional exposures, parental fertility, and nutrition on children's health, *Environmental Health Perspectives* 112.1 (2004): 69–78.

Page 62. **Caleb Finch**—C. E. Finch and R. E. Tanzi, Genetics of aging, *Science* 278: 407–411.

Page 62. **Perls points out that these findings are restricted to people living into their early 70s**—T. T. Perls et al., What does it take to live to 100? *Mechanisms of Aging and Development* 123 (2002): 231–42.

Page 63. **Richardson**—K. R. Richardson and S. H. Norgate, The equal environment assumption of classical twin studies may not hold, *British Journal of Educational Psychology* 75: 1–13.

Page 63. **Joseph**—J. Joseph, *The Gene Illusion: Genetic Research in Psychiatry and Psychology Under the Microscope* (Ross-on-Wye: PCCS, 2003).

Page 63. **Michelangelo (1475–1564) who lived to age 89**—T. T. Perls et al., Op. cit.

Page 63. **In the last half of the 1800s Denmark typically had**—G. Baggio et al., Biology and genetics of human longevity, *World Congress of Gerontology* (1997): 8–10.

Page 63. **with the number doubling every 10 years**—Ibid.

Page 64. **were seventeen times more likely to reach 100**—T. T. Perls et al., Life-long sustained mortality advantage of siblings of centenarians, *Population Biology* 99.12: 8442–8447.

Page 64. **centenarians are likely to lack a 'disease gene'**—T. T. Perls, L. M. Kunkel, and A. A., The genetics of exceptional human longevity, *Journal of American Geriatric Society* 50 (2002): 359–68.

Page 64. **from Annibale Puca**—A. Puca et al., A genome-wide scan for linkage to human exceptional longevity identifies a locus on chromosome 4, *Proc Natl Acad Sci USA* 98 (2001): 10505–10508.

Page 65. **He was also the oldest person to have a pacemaker fitted**—http://www.guinessworldrecords.com/

Page 66. **discovered that out of 157 centenarians living in Rome**—L. Tafaro et al., Smoking and longevity: An incompatible binomial, *Archives of Geront. Geriatr. Suppl.* (2004): 425–30.

Page 66. **Gary Fraser and**—G. E. Fraser and D. J. Shavlik, Ten years of life: Is it a matter of choice? *Archives of Internal Medicine* 161 (2001): 1645–1652.

Page 66. **William Cockerham and**—W. C. Cockerham, H. Hattori, and Y. Yamori, The social gradient in life expectancy: The contrary case of Okinawa in Japan, *Social Science and Medicine* 51 (2000): 115–122.

Page 66. **variation in life expectancy within**—Ibid.

Page 66. **the area having the lowest per capita**—Ibid.

Page 66. **Bradley Willcox**—B. Willcox, C. Willcox, and M. Suzuki, *The Okinawa Program: How the World's Longest-Lived People Achieve Everlasting Health—And How You Can Too* (New York: Random House, 2001).

Page 68. **The *World Cancer Report* claims that 30 per cent of**—A. Purvis, 'Eat Up Your Veg. It Could be the Next Best Thing to Giving Up Smoking', *The Observer Food Monthly*, October 2004.

Page 68. **Consuming ten portions of tomato**—E. L. Gioyannucci et al., Intake of carotenoids and retinol in relationship to risk of prostate cancer, *J. Natl Cancer Inst* 87 (1995): 1767–1177.

Page 68. **Naoko Sueoka and**—N. Sueoka et al., in Goldman et al., eds., Healthy aging for functional longevity: Molecular and cellular interactions in senescence, *Annals of New York Academy of Sciences* 928 (New York: New York Academy of Sciences, 2001): 274–79.

Page 69. **Institute of Food Research has shown that their effect can be a colossal 13 times**—A. Purvis, Op. cit., 32.

Page 69. **Heber's work**—Ibid., 28.

Page 69. **world's first pregnancy after ovarian transplant**—M. Frith, 'Ovary Transplant Success', *The Independent*, June 30, 1–2.

Page 71. **By 2025, two-thirds of the world's**—'Six Big Problems', *The Globe and Mail*, 12 October 2005.

Page 72. **examined 42 centenarians to find that the majority**—L. Tafaro et al., An investigation on behavioural problems in centenarians, *Arch. Gerontol. Geriatri. Supp. 7* (2001): 375–78.

Page 72. **Perls looked at mental clarity**—T. T. Perls, Dementia-free centenarians, *Experimental Gerontology* 39 (2004): 1587–1593.

Chapter 4

Page 74. **Erik Erikson's**—E. Erikson, *Childhood and Society* (New York: Norton, 1963), cited by M. Cole and S. R. Cole, *The Development of Children* (New York: Worth, 2001 ed.).

Pages 74–75. **Table 3**—Table based on various Web sources found by author.

Page 76. **Pia Fromholt and**—P. Fromholt et al., Life-narrative and word-cued autobiographical memories in centenarians: Comparisons with 80-year-old control, depressed, and dementia groups, *Memory* 11.1: 81–88.

Page 76. **Costa**—P. Costa and R. Kastenbaum, Some aspects of memories and ambitions in centenarians, *Journal of Genetic Psychology* 110 (1967): 3–16.

Page 77. **from Jeanne Calment**—M. Allard, V. Lebre, J. M. Robine, *Jeanne Calment from Van Gogh's Time to Ours: 122 Extraordinary Years* (New York: W. H. Freeman, 1998 ed.).

Page 77. **Michael Ross**—M. Ross and A. E. Wilson, Autobiographical memory and conceptions of self: Getting better all the time, *Current Directions in Psychological Science* 12.2 (2003): 66–69.

Page 79. **a switch between a pacemaker and accumulator**—R. Le Poidevin, A puzzle concerning the time perception, *Synthese* 142 (2004): 109–142.

Page 79. **If people see a light**—S. Droit-Volet, S. Touret, and J. Wearden, Perception of the duration of auditory and visual stimuli in children and adults, *Quarterly Journal of Experimental Psychology* 57A.5 (2004): 797–818.

Page 80. **Kuriyama**—K. Kuriyama et al., Circadian fluctuation of time perception in healthy subjects, *Neuroscience Research* 46.1 (2003): 23–31.

Page 80. **study by Ignor Nenadic**—I. Nenadic et al., Processing of temporal information and the basal ganglia: New evidence from fMRI, *Experimental Brain Research* 148 (2003): 238–46.

Page 83. **people experience déjà vu**—A. S. Brown, Getting to grips with déjà vu, *The Psychologist* 17.12: 694–96.

Page 83. **Alan Brown**—A. Brown, A review of the déjà vu experience, *Psychological Bulletin* 129 (2003): 394–413.

Page 84. **Diana Medlicott**—D. Medlicott, Surviving in the time machine: Suicidal prisoners and the pains of prison time, *Time and Society* 8.2 (1999): 211–30.

Pages 84–85. **The story of Muriel**—D. Smith, 'Living with a Terminal Illness Isn't Only a Dark Place of Despair', *The Observer*, 29 February 2004.

Page 85. **Another experience common to people facing terminal**—M. B. Chicaud, *La Crise de la Maladie Grave* (Paris: Dunod, 1998).

Page 85. **Laura Carstensen**—L. L. Carstensen and B. F. Fredricksen, Socioemotional selectivity in healthy older people and younger people living with the human immunodeficiency virus: The centrality of emotion when the future is constrained, *Health Psychology* 17 (1998): 1–10.

Page 86. **In his book**—R. Bannister, *The First Four Minutes: 50th Anniversary Edition* (Thrupp: Sutton, 2004 ed.).

Page 86. **Leonore Terr shows**—L. Terr, cited by A. Fogel, in G. Bremner and A. Slater, eds., *Theories of Development* (Oxford: Blackwell, 2004).

Page 87. **Piefke**—M. Piefke et al., Differential remoteness and emotional tone modulate the neural correlates of autobiographical memory, *Brain* 126 (2003): 650–68.

Page 88. **Rubin**—D. C. Rubin, The distribution of early childhood memories, *Memory* 8.4 (2000): 265–69.

Page 88. **Jimmy Young**—J. Young, *Forever Young: The Autobiography* (London: Hodder & Stoughton, 2004).

Page 88. **Bennett**—A. Bennett, *Writing Home* (New York: Picador, 1997 ed.).

Page 89. **Qi Wang**—Q. Wang, Infantile amnesia reconsidered: A cross-cultural analysis, *Memory* 11.1 (2003): 65–80.

Chapter 5

Page 91. **National Health Service**—S. Morris et al., Economic evaluation of strategies for managing crying and sleeping problems. *Arch. Dis . Child.* 84: 15–19.

Page 92. **A fetus can spend up 80 per cent of its time asleep**—W. Dement and C.Vaughan, *The Promise of Sleep: The Scientific Connection Between Health, Happiness, and a Good Night's Sleep* (New York: Macmillan, 1999), 104.

Page 92. **heart rate can be detected**—M. Serron-Ferre et al., The development of circadian rhythms in the fetus and neonate, *Semin. Perinatol.* 25 (2001): 363–70.

Page 92. **Even at 17 weeks after birth**—S. W. D'Souza et al., Skin temperature and heart Rate rhythms in infants of extreme prematurity, *Arch. Dis. Child.* 67 (Spec. No.): 784–88, cited by S. A. Rivkees, Developing circadian rhythmicity in infants, *Pediatrics* 112.2 (2003): 373–81.

Page 92. **not detectable in babies until 2 months**—S. A. Rivkees, Developing circadian rhythmicity in infants, *Pediatrics* 112.2 (2003): 373–81.

Page 92. **have daily oscillations *in utero***—H. Hao and S. A. Rivkees, The biological clock of very premature primate infants is responsive to light, *Proc Natl Acad Sci. USA* 96: 2426–2429.

Page 92. **is present at least 36 weeks after conception, if not earlier**—S. F. Glotzbach, P. Sollars, M. Pagano, Development of the human retinohypothalmic tract, *Soc Neurosci.* 18 (1992): 857, cited by S. A. Rivkees, Developing circadian rhythmicity in infants, *Pediatrics* 112.2 (2003): 373 81.

Page 92. **Scott Rivkees**—S. A. Rivkees et al., Rest-activity patterns of premature infants are regulated by cycled lighting, *Pediatrics* 113.4 (2004): 833–39.

Page 93. **Yvonne Harrison**—Y. Harrison, The relationship between daytime exposure to light and night-time sleep in 6–12-week-old infants, *J. Sleep Research* 13: 345–52.

Page 93. **Sari Goldstein-Ferber**—S. Goldstein-Ferber, Massage therapy by mothers enhances the adjustment of circadian rhythms to the nocturnal period in full-term infants, *Developmental and Behavioural Pediatrics* 23.6 (2002): 410–15.

Page 95. **DeCasper**—A. J. DeCasper and M. J. Spence, Prenatal maternal speech influences newborns' perception of speech sounds, *Infant Behaviour and Development* 9 (1986): 133–50.

Page 95. **Franck Ramus**—F. Ramus et al., Language discrimination by human newborns and by Cotton-top Tamarin monkeys, *Science* 288: 349–51.

Page 96. **Cynthia Crown**—C. L. Crown, Coordinated and interpersonal timing of vision and voice as a function of interpersonal attraction, *Journal of Language and Social Psychology* 10.1 (1991): 29–46.

Page 96. the duration of—J. Jaffe, Rhythms of dialogue in infancy: Coordinated timing in development, *Monographs for the Society for Research in Child Development*, Serial no. 265, vol. 66, no. 2 (2001): 1.

Page 96. **An adult talking with a stranger**—J. Jaffe, ibid.

Page 97. **Joseph Jaffe**—J. Jaffe, ibid.

Page 97. **Daniel Stern**—D. Stern, Commentary: Rhythms of dialogue in infancy: Coordinated timing in development, *Monographs for the Society for Research in Child Development*, Serial no. 265, vol. 66, no. 2 (2001): 144.

Page 97. **John Colombo**—J. Colombo and W. A. Richman, Infant timekeeping: Attention and temporal estimation in 4-month-olds, *Psychological Science* 13.5 (2002): 475–79.

Page 98. **Carolyn Rovee-Collier**—C. Rovee-Collier, The development of infant memory, *Current Directions in Psychological Science* 8.3: 80–83.

Page 99. **Jean Piaget**—J. Piaget, *The Construction of Reality in the Child* (London: Routledge, 1954), cited by D. Kuhn and R. S. Siegler (vol. eds.), *Handbook of Child Psychology*, vol. 2 (New York: John Wiley), 220.

Page 99. **Peter Willats**—Cited by D. Kuhn and R. S. Siegler, ibid.

Page 100. **David Lewkowicz**—D. J. Lewkowicz, Perception of serial order in infants, *Developmental Science* 7.2 (2004): 175–84.

Page 100. **Babies aged 5-months**—C. von Hofsten and L. Ronnqvist, Preparation for grasping an object: A developmental study, *Journal of Experimental Psychology: Human Perception and Performance* 14.4: 610–21.

Chapter 6

Page 105. **Benjamin Libet**—B. Libet, *Mind-Time: The Temporal Factor in Consciousness* (Cambridge: Harvard University Press, 2004).

Page 105. **Irwin Lee**—I. H. Lee and J. A. Assad, Putaminal activity for simple reactions or self-timed movements, *J. Neurophysiol* 89: 2528–2537.

Page 107. **Michael Land**—M. F. Land and P. McLeod, From eye movements to actions: How batsmen hit the ball, *Nature Neuroscience* 3.12 (2000): 1340–1345.

Page 108. **Fine and Donald Macleod from Stanford University and the Salk Institute in California met Michael May**—http://www.webmd.com

Page 108. **Karl Becker**—K. Becker, A. A. Raheja, and C. Woody, A glimpse of the brain transforming a sensory signal into a motor response, *Somatosensory & Motor Response* 19.4 (2002): 296–301.

Page 109. **Muscles trained for intense short bursts**—T. M. Williams et al., Skeletal muscle histology and biochemistry of an elite sprinter, the African cheetah, *J. Comp. Physiol B* 167 (1997): 527–35.

Page 109. **capable of accelerating to over 60 mph in less than three seconds**—Ibid.

Page 109. **Wild cheetahs have as much as 82 per cent of**—Ibid.

Page 110. **Ricarda Schubotz**—R. I. Schubotz et al., Time perception and motor timing: A common cortical and subcortical basis revealed by fMRI, *NeuroImage* 11 (2000): 1–12.

Page 110. **Meegan**—D. V. Meegan, R. N. Aslin, and R. A. Jacobs, Motor timing learned without motor training, *Nature Neuroscience* 3.9 (2000): 860–62.

Page 112. **Pollok**—B. Pollok et al., Cortical activations associated with auditorily paced finger tapping, *Neuroreport* 14.2 (2003): 247–50.

Page 112. **Researchers from the Otto-von-Guericke University of Magdeburg**—L. Jäncke et al., Cortical activations during paced finger-tapping: Applying visual and auditory pacing stimuli, *Cognitive Brain Research* 10 (2000): 51–66.

Page 112. **Claire Calmels**—C. Calmels and J. F. Fournier, Duration of physical and mental execution of gymnastic routines, *The Sport Psychologist* 15 (2001): 142–50.

Chapter 7

Page 116. **Hillel**—A. Hillel, The study of laryngeal muscle activity in normal human subjects and in patients with laryngeal dystonia using multiple fine-wire electromyography, *Laryngoscope* 111 (2001).

Page 117. **Chara Malapani**—C. Malapani et al., Coupled temporal memories in Parkinson's disease: A dopamine-related dysfunction, *Journal of Cognitive Neuroscience* 10 (1998): 316–31.

Page 118. **Jon Hore**—J. Hore, D. Timmann, and S. Watts, Disorders in timing and force of finger opening in overarm throws made by cerebellar subjects, *Annals of the New York Academy of Science* 978 (2002): 1–15.

Page 119. **Anna Smith**—A. Smith et al., Evidence for a pure time perception deficit in children with ADHD, *Journal of Child Psychology and Psychiatry* 43.4 (2002): 529–42.

Page 120. **by an average of £20,688 per year**—http:// prnewswire.co.uk/cgi/news/ release?id = 23546

Page 120. **the disease currently affects 19 per cent of people**—Alzheimer's Association, *Alzheimer's Disease Fact Sheet* (Chicago: Author, 1996), cited by G. White and A. C. Ruske, Memory deficits in Alzheimer's disease: The encoding hypothesis and cholingergic Function, *Psychonomic Bulletin and Review* 9.3: 426–37.

Page 120. **Damien Leger**—D. Leger et al., Sleep/wake cycles in the dark: Sleep recorded by polysomnography in 26 totally blind subjects compared to controls, *Clinical Neurophysiology* 113.10 (2002): 1607–1614.

Page 121. **babies born blind and deaf**—G. Medicus, M. Schleidt, and I. Eibleibesfeldt, Universal time-constant in movements of deaf and blind children, *Nervenarzt* 65.9 (1994): 598–601.

Page 121. **John Hull**—J. Hull, *Touching the Rock: An Experience of Blindness* (London: Arrow, 1990), cited by O. Sacks, *An Anthropologist on Mars* (London: Picador, 1995).

Page 122. **My own research** –S. Norgate, G. M. Collis, V. Lewis, The developmental role of rhymes and routines for congenitally blind children, *Cahiers de Psychologie Cognitive /Current Psychology of Cognition* 17.2 (1998): 451–79.

Page 122. **Haddon's**—M. Haddon, *The Curious Incident of the Dog in the Night-Time* (London: Vintage, 2004 ed.).

Page 123. **cue in the environment until some 800–1200 milliseconds**—D. Wimpory, Social timing, clock genes and autism: A new hypothesis, *Journal of Intellectual Disability Research* 46.4 (2002): 352–58.

Page 123. **Although melatonin levels are in phase**—Nir et al., Brief report: Circadian melatonin, thyroid-stimulating hormone, prolactin, and cortisol levels in serum of young adults with autism, *Journal of Autism and Developmental Disorders* 25.6 (1995).

Page 124. **Special talent in calendar calculating**—L. K. Miller, The savant syndrome: Intellectual impairment and exceptional skill, *Psychological Bulletin* 125.1 (1999): 31–46.

Page 124. **Pring**—L. Pring and B. Hermelin, Numbers and letters: Exploring an autistic savant's unpractised ability, *Neurocase* 8 (2002): 330–37.

Page 125. **Jill Boucher**—J. Boucher, in C. Hoerl and T. McCormack, eds., *Time and Memory: Issues in Philosophy and Psychology* (Oxford: Clarendon Press, 2001), 111–35.

Page 125. **Lapp**—Lapp et al., Psychopharmacological effects of alcohol on time perception: The extended balanced placebo design, *Journal of Studies on Alcohol* (1994): 96–12.

Page 125. **Tara Wass**—T. S. Wass et al., Timing accuracy and variability in children with prenatal exposure to alcohol, *Alcoholism: Clinical and Experimental Research* 26.12 (2002): 1887–1896.

Page 126. **Matthias Brand**—M. Brand et al., Cognitive estimation and affective judgements in alcoholic Korsakoff patients, *Journal of Clinical and Experimental Neuropsychology* 25.1 (2003): 324–34.

Page 126. **Philip Zimbardo**—P. G. Zimbardo and J. N. Boyd, Putting time in perspective: A valid, reliable individual-differences metric, *Journal of Personality and Social Psychology* 77.6 (1999): 1271–1288.

Page 127. **Keough**—Keough et al., Who's smoking, drinking, and using drugs? Time perspective as a predictor of substance use, *Basic and Applied Social Psychology* 21.2: 149–64.

Page 127. **Nancy Petry**—N. Petry, W. K. Bickel, and M. Arnett, Shortened time horizons and insensitivity to future consequences in heroin addicts, *Addiction* 93.5 (1998): 729–38.

Chapter 8

Page 133. **certain DNA strand and protein damage**—R. Chapin et al., Op. cit.

Page 133. **lower sperm counts across as many as four subsequent generations**—S. Connor, 'Man-made Pesticides Blamed for Fall in Male Fertility Over Past 50 Years', *The Independent*, 3 June 2005, 20.

Page 133. **pre-natal exposure to alcohol**—T. S. Wass et al., Op. cit.

Notes

Page 133. **communicating with a stranger, would be matched**—C. L. Crown, Coordinated and interpersonal timing of vision and voice as a function of interpersonal attraction, *Journal of Language and Social Psychology* 10.1 (1991): 29–46.

Page 134. **you will be living at a faster pace than in Rio de Janeiro**—R. V. Levine and A. Norenzayan, Op. cit.

Page 135. **newborns can differentiate**—F. Ramus et al., Language discrimination by human newborns and by Cotton-top Tamarin monkeys, *Science* 288: 349–51.

Page 135. **autobiographical memory may extend back earlier**—Q. Wang, Op. cit.

Page 135. **The Nuer of Sudan**—F. Fernández Armesto, *Ideas That Changed the World* (London: Dorling Kindersley, 2003), 112.

Page 136. **Nunez**—*The Guardian*, 24 February 2005, 9.

Page 136. **In Japan, teenagers playing**—M. Maruyama, The new logic of Japan's young generation, technological forecasting and social change, *Annals of Internal Medicine* 28 (1985): 351–64.

Page 137. **usually a delay of a few months until**—S. A. Rivkees, Developing circadian rhythmicity in infants, *Pediatrics* 112.2 (2003): 373–81.

Page 138. **In adolescence there is a phase delay**—F. Giannotti et al., Circadian preference, sleep and daytime behaviour in adolescence, *J. Sleep Res.* 11 (2002): 191–99.

Page 138. **with age the crystalline lens of the eye**—W. N. Charman, Age, lens transmittance, and the possible effects of light on melatonin suppression, *Ophthal. Physiol.* 23 (2003): 181–87.

Page 139. **one billion people live on the equivalent of a dollar a day**—J. Vandemoortele, in P. Townsend and D. Gordon, eds., *World Poverty: New Policies to Defeat an Old Enemy* (Southampton: Policy Press, 2004 ed.), 378.

Page 139. **pet owners currently spend around $11.6 billion**—D. Gordon cites Euromonitor International (2001) in P. Townsend and D. Gordon, eds., *World Poverty: New Policies to Defeat an Old Enemy*, 73.

Page 139. **average weight of American adults has increased by 25 pounds since 1960**—N. Hellmich, 'Average Weight up 25 lbs. Since 1960', *USA Today*, 28 October 2004.

Page 139. **only 25 countries in the world have a larger**—P. Townsend, in P. Townsend and D. Gordon, eds., Op. cit., 11.

Page 139. **The African Commission of Human and Peoples' Rights**—K. Wiwa, 'It Just Won't Work', *The Observer*, 12 June 2005.

Page 139. **the case of Ghana, where one-third of the population still depends on untreated spring or river water**—K. Donkor, in P. Townsend and D. Gordon, eds., Op. cit., 221.

Page 139. **53 per cent of heads of household have never spent any time in education**—Rothchild (1991), cited in K. Donkor, in P. Townsend and D. Gordon, eds., Op. cit., 220.

Pages 139–40. **61 per cent have access to health services**—K. Donkor, in P. Townsend and D. Gordon, eds., Op. cit., 219.

169

Page 140. **Africa sends 77,000 professionals abroad**—K. Wiwa, Op. cit.

Page 140. **Nancy Petry is relevant**—N. Petry, W. K. Bickel, and M. Arnett, Shortened time horizons and insensitivity to future consequences in heroin addicts, *Addiction* 93.5 (1998): 729–38.

Page 141. **coronary heart disease by 40 per cent**—H. Boggild, A. Knutsson, Shiftwork, risk factors and cardiovascular disease (reviews), *Scand. J. Work Environ. Health* 25.2 (1999): 85–99.

Page 141. **The British Heart Foundation**—http://www.bhf.org.uk/professionals

Page 141. **one in seven people in the UK work between 6 p.m. and 9 a.m.**—http://scotsman.com/index.cfm?id = 1133572004

Page 141. **As for the social impact, 58 per cent of people questioned by the Future Foundation**—http://scotsman.com/index.cfm?id = 1133572004

Page 141. **progress on under-five mortality was slower in the last decade than in any since 1960**—J. Vandemoortele, P. in Townsend and D. Gordon, eds., Op. cit., 379.

Page 141. **Some 120 million children spend no time in primary-level education**—Ibid.

Page 141. **average spent on education amounted to less than 4 per cent of the gross domestic product in sub-Saharan Africa**—in P. Townsend and D. Gordon, eds., Op. cit., 221.

Page 142. **For the 300,000 children worldwide who are soldiers**—A. H. Hoiskar, Under age and under fire: An enquiry into the use of child soldiers, 1994–98, *Childhood* 8.3 (2001): 340.

Page 142. **there are 100,000 blind children**—http://www.who.int/inf-pr-2000–09.html

Page 142. **avoidable blindness by 2020**—http://www.who.int/inf-pr-2000–09.html

Page 142. **350 million women**—S. George, Op. cit.

Page 143. **Working mothers**—M. Frith, 'Working Mothers Suffer Record Levels of Sleep Deprivation', *The Independent*, 2 June 2005.

Page 143. **one in four women over 50**—M. Hughes, 'Giving Home Carers Chance of Life Outside', *The Guardian*, 18 June 2005.

Page 143. **UK government**—Home Office, *Supporting Families: A Consultation Document* (London: HMSO, 1998).

Page 144. **Pia Christensen**—P. H. Christensen, Why more 'quality time' is not on the top of children's lists: The 'qualities of time' for children, *Children and Society* 16 (2002): 77–78.

Page 145. **Japanese government**—J. McCurry, J., 'How Japan grew bored with love', *The Observer*, 5 June 2005.

Page 145. **Work Foundation**—http://www.employersforwork-lifebalance.org.uk

Page 145. **BT**—Ibid.

Page 146. **immersed in 'flow'**—M. Csikszentmihalyi, *Flow: The Psychology of Happiness* (London: Rider, 1992).

Page 146. **Given that in European countries, three-quarters of employees work outside the traditional 9 to 5**—G. Costa et al., Op. cit.

Page 146. **It was only 150 years ago that wristwatches first came into use**—R. Levine, Op. cit.

Page 147. **120 years ago when strong artificial light became widely available domestically**—L. Kreitzman, Op. cit., 23.

Page 147. **aviation industry is set to rise by 5–7 per cent per year**—Hume et al., Complaints caused by aircraft operations: An assessment of annoyance by noise level and time of day, *Journal of Air Transport Management* 9 (2003): 153–60.

Page 148. **incidences of malignant melanoma have been found to be two to three times higher**—http://www.hon.ch/News/HSN/515697.html

Page 148. **two-thirds of British households experience night-time noise levels higher**—L. Mangan, 'Peace Struggle', *The Guardian*, 25 January 2005, 2.

Page 149. **Carol Saunders**—C. Saunders, Van Slyke, and D. R. Vogel, My time or yours? Managing time visions in global virtual teams, *Academy of Management Executive* 18.1 (2004): 19–31.

Page 150. **tourist trips to Mars**—T. Watson, T., 'A Lot of Ground to Be Covered Before Space Tourism Can Fly', *USA Today*, 29 October 2004.

Page 151. **overseen by the National Physical Laboratories atomic clock**—Advert in *The Independent* on Sunday, 10 April 2005, 52.

BIBLIOGRAPHY

Adam, B., *Time and Social Theory* (Philadelphia: Temple University Press, 1990).

Allard, M., Lebre, V., Robine, J.M., *Jeanne Calment From Van Gogh's Time to Ours: 122 Extraordinary Years* (New York: W.H. Freeman and Company, 1998 edition).

Bannister, R., *The First Four Minutes: 50ᵗʰ Anniversary Edition* (Thrupp: Sutton Publishing, 2004 edition).

Bunting, M., *Willing Slaves: How the Overwork Culture is Ruling Our Lives* (London: Harper Perennial, 2004).

Clark, W.R., *A Means to an End: The Biological Basis of Aging and Death* (Oxford: Oxford University Press, 1999).

Csikszentmihalyi, M., *Flow: The Psychology of Happiness* (London: Rider, 1992).

Dement, W., and Vaughan, C., *The Promise of Sleep: The Scientific Connection Between Health, Happiness, and a Good Night's Sleep* (New York: Macmillan, 1999).

Eriksen, T.H., *Tyranny of the Moment: Fast and Slow Time in the Information Age* (London: Pluto Press, 2001).

Finch, C.E., and Kirkwood, T.B.L., *Chance, Development, and Aging* (Oxford: Oxford University Press, 2000).

Foster, R.G., and Kreitzman, L., *Rhythms of Life: The Biological Clocks that Control the Daily Lives of Every Living Thing* (London: Profile Books, 2004).

Gershuny, J., *Changing Times: Work and Leisure in Postindustrial Society* (Oxford: Oxford University Press, 2000).

Gleick, J., *Faster: The Acceleration of Just About Everything* (London: Little, Brown and Company, 1999).

Griffiths, J., *A Sideways Look at Time* (New York: Jeremy P. Tarcher, 2002 edition).

Honoré, C., *In Praise of Slow: How a Worldwide Movement is Challenging the Cult of Speed* (London: Orion, 2004).

Kreitzman, L., *The 24 Hour Society* (London: Profile Books, 1999).

Levine, R., *A Geography of Time: The Temporal Misadventures of a Social Psychologist, or How Every Culture Keeps Time Just a Little Bit Differently* (New York: Basic Books, 1997).

Olshansky, S.J., and Carnes, B.A., *The Quest for Immortality: Science at the Frontiers of Aging* (New York: W.W. Norton and Company, 2001).

Ornstein, R.E., *On the Experience of Time* (Harmondsworth: Penguin Books, 1969).

Overall, C., *Ageing, Death and Human Longevity: A Philosophical Inquiry* (Berkeley: University of California Press, 2003).

Perls, T.T., and Silver, M.H., *Living to 100: Lessons in Living to Your Maximum Potential at Any Age* (New York: Basic Books, 1999).

Robinson, J.P., and Godbey, G., *Time for Life: The Surprising Ways Americans Use Their Time* (Pennsylvania: Pennsylvania State University Press, 1997).

Segal, A., (ed.) *Molecular Biology of Circadian Rhythm*, (New Jersey: John Wiley, 2004).

Tomlinson, J., and Tomlinson, M., *The Luxury of Time* (New York: Pocket Books, 2005).

Townsend, P., and Gordon, D., (eds) *World Poverty: New Policies To Defeat An Old Enemy* (Southampton: The Policy Press, 2004 edition).

Willcox, B., Willcox, C., and Suzuki, M., *The Okinawa Program: How The World's Longest-Lived People Achieve Everlasting Health – And How You Can Too* (New York: Random House, 2001).

Williams, A.M., Davids, K., and Williams, J.G., *Visual Perception and Action in Sport* (London: E & FN Spon, 1999).

INTERNET RESOURCES

A UK charity dedicated to the provision of safe domestic water, sanitation and hygiene promotion to the world's poorest people.
www.wateraid.org

The National Autistic Society
http://www.nas.org.uk/

Saving lives of children affected with eye cancer.
http://www.thedaisyfund.org/

Childhood Eye Cancer Trust
http://www.chect.org.uk/

Royal National Institute for the Blind
http://www.rnib.org.uk

American Foundation for the Blind
http://www.afb.org/

Control of Aircraft Noise
http://www.scan-uk.mmu.ac.uk

Heathrow Association for the Control of Aircraft Noise
http://www.hacan.org.uk/

Marie Stopes International provides access to sexual and reproductive health facilities.
http://www.mariestopes.org.uk/

British Association for Fair Trade Shops, working to stop exploitative child labour.
http://www.bafts.org.uk/

An international humanitarian organisation, Plan helps children to realise their full potential by helping to fight poverty.
http://www.plan-uk.org/

ActionAid works in Africa, Asia, Latin America and the Caribbean, listening to, learning from and working in partnership with over nine million of the world's poorest people.
www.actionaid.org.uk

CARE International is a global humanitarian organisation, working with over thirty million disadvantaged people each year in seventy-two of the world's poorest countries.
www.careinternational.org.uk

Oxfam GB is a development, relief and campaigning organisation that works with others to find lasting solutions to poverty and suffering around the world.
www.oxfam.org.uk

Working to improve the quality of carers' lives.
www.carersuk.org

Alzheimer's Association
http://www.alz.org/

The Work Foundation exists to inspire and deliver improvements to performance through improving the quality of working life.
http://www.theworkfoundation.com/index.jsp

London 2012 Olympic Games
http://www.london2012.org/en

UK Athletics Club Website Directory
http://www.runtrackdir.com/ukclubs/

Slow Food – International movement opposing fast food and promoting dining as a source of pleasure.
http://www.slowfood.com/

Sweatshop watch – serves low-wage workers nationally and globally, with a focus on eliminating sweatshop exploitation in California's garment industry.
http://www.sweatshopwatch.org/

Stopping the use of child soldiers: Human Rights Watch – defending human rights worldwide.
http://hrw.org/

A charity which works primarily using the services of volunteers.
http://vso.org.uk/

Time-saving gadgets
http://www.innovations.co.uk

INDEX

Index